D1323570

Sound Engineering
Explained

Sound Engineering Explained

Second Edition

Michael Talbot-Smith

BSc, CPhys, MinstP

ELSEVIER

AMSTERDAM • BOSTON • HEIDELBERG • LONDON • NEW YORK •OXFORD
PARIS • SAN DIEGO • SAN FRANCISCO • SINGAPORE • SYDNEY • TOKYO

Focal Press is an imprint of Elsevier

Focal Press
An imprint of Elsevier
Linacre House, Jordan Hill, Oxford OX2 8DP
30 Corporate Drive, Burlington, MA 01803

First published as Audio Explained 1997
Reprinted 1998
Second edition published as Sound Engineering Explained 2002
Reprinted 2002, 2004, 2005

British Library Cataloguing in Publication Data
A catalogue record for this book is available from the British Library

Library of Congress Cataloguing in Publication Data
A catalogue record for this book is available from the Library of Congress

ISBN 0 240 51667 2

> For information on all Focal Press publications
> visit our website at www.focalpress.com

Working together to grow
libraries in developing countries

www.elsevier.com | www.bookaid.org | www.sabre.org

ELSEVIER   BOOK AID International   Sabre Foundation

Printed and bound in Great Britain by Biddles Ltd, King's Lynn, Norfolk

Contents

Preface   xi

Acknowledgements   xiii

About this book   xv

1   **Sound waves**   **1**
    Part 1   **1**
    What are sound waves?   1
    Frequency   2
    Wavelength   2
    Amplitude   2
    The velocity of sound waves   3
    Velocity, frequency and wavelength   3
    Sound waves and obstacles   4
    The bending of sound waves (diffraction)   5
    Units used in sound   6
    Sound intensity and the effect of distance   9
    Decibels   10

    Part 2   **11**
    The velocity of sound waves   11
    Units   12
    Decibels   12

2   **Hearing and the nature of sound**   **14**
    Part 1   **14**
    The response of the ear   14
    Loudness. The dB(A)   15
    Musical pitch   16

Musical quality ('timbre')   18
The brain's perception of sound   19

Part 2   22
Pitch   22
Frequency ranges in music   22
'False bass'   23

3   **Basic acoustics**   25
    Part 1   25
    Sound isolation   25
    Internal acoustics   29

    Part 2   35
    Sound isolation   35
    Sound absorption   35

4   **Microphones**   40
    Part 1   40
    Microphone transducers   40
    Polar responses   44
    'Boundary' microphones   48
    Personal microphones   48
    Radio microphones ('wireless microphones')   48

    Part 2   51
    Electrostatic microphones   51
    Production of the different polar responses   52
    Sensitivities of microphones   56
    Phantom power   57
    Balanced wiring   58
    Radio microphone data   59

5   **Using microphones**   61
    Objectives and problems in recording   61
    Specific applications of microphones   66

6   **Monitoring**   71
    Part 1   71
    Technical monitoring   71
    Aural monitoring   75
    Listening tests   80

Part 2   **82**
Reference voltages in audio signals   82
PPMs and VU meter readings   83
Helmholtz resonators   84
Loudspeaker power   85
Loudspeaker cables   86
Professional and domestic standards   87

7   **Stereo**   **88**
    Part 1   **88**
    How stereo works   88
    Methods of producing inter-channel differences   90
    Terminology   91
    Stereo listening   92
    Stereo loudspeaker matching   94
    Phase   94

    Part 2   **97**
    Microphone techniques for stereo   97
    Headphones for stereo monitoring   102

8   **Sound mixers**   **104**
    Part 1   **104**
    Terminology   1005
    The basic channel   105
    Important features of any mixer   107
    Output stage   111

    Part 2   **113**
    Inputs and connections   113
    Equalization   116
    Public Address and Foldback (PA and FB)   119
    Talkback   121

9   **Controlling levels**   **123**
    Manual control of levels   123
    Electronic level control   127
    Noise gates   131

10 **Digital audio**   **133**
    Part 1   **133**
    Historical   133
    Basic principles   133
    Other applications of digital audio   139

MIDI   142
Data compression   143
Additional terminology   143

Part 2   **145**
Compact discs   145
Error correction   147
Cleaning CDs   148
Cost of CD players   148
NICAM   148
A little more about error correction   150

11 Recording   **151**
Part 1 – Recording devices and systems   **151**
Analogue recording   151
The fundamentals of magnetic recording   152
Cassette quality   153
Noise reduction   154
Head and tape cleanliness   155
Digital recording   156
DAT   157
MiniDisc®   157
Solid state recording   158
MP3   159

Part 2 – Editing   **160**
Why edit?   160
Practicalities   161
Editing tape   162
Digital tape editing   162
MiniDisc editing   163

12 Public address   **164**
Part 1   **164**
Outdoor PA   165
Indoor PA   165

Part 2   **167**
Howl-rounds   167
100-volt systems   168

13 Music and sound effects 170
Music 170
Sound effects 171

14 Safety 175
Electrical safety 175
Noise and hearing 178
'Mechanical' safety 179
Fire 181

Copyright 184

Miscellaneous data 185

Further reading 189

Answers 191

Index 195

Preface

This book has had two previous incarnations. It first appeared in 1994 under the heading of *Audio Recording and Reproduction: Practical Measures for Audio Enthusiasts.* Gratifyingly, only three years later, there was a need for a second edition and the publishers, feeling quite rightly that the original title was rather cumbersome, suggested a new title and a revision of the whole book. It then appeared as *Audio Explained.*

The aim of the original was to make it easier for people without professional equipment to make recordings of a very respectable standard for whatever purpose – I suggested conductors of amateur choirs or orchestras, teachers who needed to develop their own audio-visual material, members of drama groups, and so on. I also had in mind for the potential readership the large number of people who are interested in how their hi-fi equipment works and also those with camcorders who want to edit their work and enhance it with slightly better and possibly more ambitious sound!

Then, more recently, it appeared that *Audio Explained* came near to being appropriate for students doing the City and Guilds Sound Engineering 1820 Course. So, it's been thoroughly revised, updated and generally (I hope) improved so that it caters for these students as well.

I'd like to think, though, that it's still useful to the original targets.

I've worked throughout on my very firmly held conviction that a degree of understanding of the basic principles is essential. To help do that and still make the book reasonably readable, most chapters are in two parts. The first part of a chapter should be easy reading for any interested person, no matter how slight their previous knowledge. The second part deals with the topic in a little more depth, but again no great technical knowledge is needed.

The intentions therefore are threefold:

1. To be a useful guide to the non-professional practitioner
2. To explain as far as possible in a non-technical way how modern audio works
3. To be useful to serious students of sound technology.

Michael Talbot-Smith

Acknowledgements

As I have explained in the Preface, this is the third version of a book which appeared first in 1994, and that means it's generally been successful. So who gets the credit? I'd like to think I do, but honesty makes me say that there are others who've made very useful contributions.

First was an old friend, Geoff Atkins, one-time Senior Audio Supervisor in BBC Wales and currently a consultant and freelance sound supervisor. He went through the original draft and made many suggestions which improved my work very considerably. Also, Eric Carter, a senior member of the staff at the Cable and Wireless Training College in Coventry, was very helpful with comments about Safety.

More recently, John Mizzi, Chief Examiner for the City and Guilds Sound Engineering 1820 course, read through the manuscript and suggested how it might be improved, so there's a third person to whom I'm grateful.

But finally I must acknowledge with great pleasure the help and encouragement provided by Beth Howard, Associate Editor at Focal Press. She's understanding whenever I have had problems and I've always felt that she's a friend at the other end of the phone line (or the e-mail link!).

Michael Talbot-Smith

About this book

The route from performer to listener

Bold print items are among those dealt with in the book.

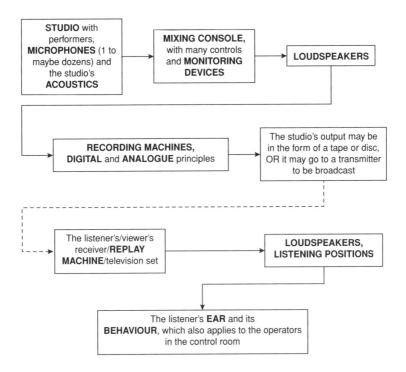

STUDIO with performers, **MICROPHONES** (1 to maybe dozens) and the studio's **ACOUSTICS**

MIXING CONSOLE, with many controls and **MONITORING DEVICES**

LOUDSPEAKERS

RECORDING MACHINES, DIGITAL and **ANALOGUE** principles

The studio's output may be in the form of a tape or disc, OR it may go to a transmitter to be broadcast

The listener's/viewer's receiver/**REPLAY MACHINE**/television set

LOUDSPEAKERS, LISTENING POSITIONS

The listener's **EAR** and its **BEHAVIOUR**, which also applies to the operators in the control room

In this book definitions and items of information which may be particularly useful are put into boxes with a single line, like this one.

Tinted boxes contain more advanced material which is not essential but some readers may find interesting or useful.

1 Sound waves
Part 1

Some initial knowledge of sound waves is essential if later chapters are going to be fully meaningful. The following is a brief outline of their most important properties. Scientific facts and figures are set out in Part 2 at the end of this chapter for those who may find them useful or interesting.

What are sound waves?

They are described as waves of compression and rarefaction in the air. This means that when sound waves travel past a fixed point the atmospheric pressure at that point goes slightly above and below the steady barometric pressure. But these fluctuations are far too small and far too rapid ever to be registered by a barometer. A microphone, on the other hand, can be thought of as a kind of extra-sensitive electrical barometer capable of detecting these rapid fluctuations in the air pressure, unlike a normal barometer, which can only indicate relatively slow pressure changes. Figure 1.1 shows what a sound wave would look like if we could see it.

DEFINITIONS
1. *Compression* – a region where the air is compressed. In sound waves the compression is very small indeed.
2. *Rarefaction* – the opposite of a compression. The air pressure is slightly lower than normal.
3. '*Steady barometric pressure*' – the normal air pressure of the atmosphere.

There is a *compression* where the lines are closest together and a *rarefaction* where they are widely spaced. Waves of this sort are called *longitudinal waves*. The waves we see in water, for instance, are

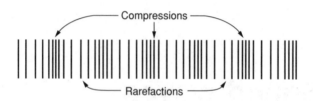

Figure 1.1 Compression and rarefactions

called *transverse waves* because the individual bits (molecules) of the water move up and down when the wave goes horizontally. In longitudinal waves, like sound, the air molecules oscillate from side to side.

Despite the fact that sound waves are longitudinal, we always draw them like *transverse waves* – it's much easier!

Frequency

This is the number of waves emitted, or received, in one second. Up to the 1950s, frequency was expressed in English-speaking countries as so many *cycles per second* (*c.p.s.* or c/s), but this has been replaced by the *hertz*, to commemorate the name of the German physicist, Heinrich Hertz (1857–1894), who was associated with early work in radio waves. The abbreviation is 'Hz'. This unit is commonly encountered in the home on mains-powered equipment, where the specified frequency of the supply in most of Europe is 50 Hz.

The normal human ear can detect sound wave frequencies in the range from approximately 16 Hz to about 16 kHz (kHz = kilohertz, or thousand hertz – 16 kHz is 16 000 Hz). The ear's response to sound will be dealt with more fully in Chapter 2.

Related to frequency is *period* – the time duration of one cycle. It is equal to $1/f$.

Wavelength

With any wave the distance between corresponding points on successive waves is termed the *wavelength*. The symbol used for wavelength is the Greek letter λ (pronounced 'lambda'). Figure 1.2 illustrates wavelength.

Amplitude

This is the 'height' of the wave in whatever units are most convenient – in electrical work, for example, the amplitude could be expressed in volts or amperes.

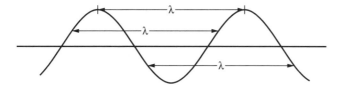

Figure 1.2 Wavelength

The velocity of sound waves

In the air, under normal conditions, sound waves travel at about 340 metres per second (m/s). This is close to 760 miles/hour, or about 1120 feet/second. The velocity varies slightly with air temperature, which is why players of wind instruments need, literally, to warm up their instruments, as the pitch depends on the speed with which sound waves oscillate within the instrument. For most practical purposes, apart from the instance just quoted, the variation in sound wave velocity with temperature is unimportant. Details are given in Part 2 of this chapter. The symbol for sound wave velocity is, in this book, v.

> Sound waves travel in air at about 340 m/s, i.e. $v = 340$ m/s.

Velocity, frequency and wavelength

These three things are linked by an important, but fortunately simple, formula. A moment's thought will show that if, say, 1000 waves per second are being emitted from a source (i.e. the frequency is 1 kHz) and each has a wavelength of λ, then after one second the first waves emitted will be $1000 \times \lambda$ away. Or, their velocity is $1000 \times \lambda$ metres per second.

What we've done is to multiply the frequency by the wavelength to arrive at the velocity:

$$\text{Frequency} \times \text{Wavelength} = \text{Velocity (or } f \times \lambda = v)$$

> IMPORTANT
> Frequency $\times$ Wavelength = Velocity (or $f \times \lambda = v$)

Writing this another way:

$$\text{Wavelength} = \text{Velocity/Frequency (or } \lambda = v/f)$$

Putting 340 m/s as the velocity and 1000 Hz as the frequency we have

Wavelength = 340/1000 metres
= 0.34 metres (or 34 cm)

It should be fairly obvious that if the frequency is doubled (to 2 kHz) then the wavelength is halved (to 17 cm) and so on. Simple calculations along these lines will show that the lowest frequency the normal ear can detect – 16 Hz – corresponds to a wavelength of about 21 m (70 feet), while at the highest frequency, around 16 kHz, the wavelength is about 2 cm, or rather less than one inch.

We shall see in the next two sections that this vast range of wavelengths presents problems in many sound recording situations, as the degree to which sound waves are reflected or bent round obstacles depends critically on the wavelength.

Sound waves and obstacles

Everyone is familiar with echoes – sounds being reflected from a large building or a cliff. What is perhaps not so obvious is that for this reflection to occur to any significant effect the sound wavelength must be smaller than the dimensions of the reflecting object. For example, the side of a building might be 10 m high and 20 m long. Sounds striking this building at right angles will be reflected if their wavelength is less than 10 m – that is, for frequencies greater than about 30 Hz. The most predominant frequencies in the human voice range from roughly 200 Hz to around 500 Hz, so a shout near to this building will be easily reflected.

On the other hand, if the object has dimensions less than wavelength there is no reflection. Instead the waves bend round the object, a process called *diffraction*, which we shall deal with in the next section.

If sound waves with a short wavelength strike an object and are reflected, there is what is termed a *sound shadow* behind the object. Figure 1.3 shows what happens in those cases when the wavelength is greater (a) and smaller (b) than the obstacle.

There is no sharp division between reflection and diffraction. For example, and taking simple figures, suppose we have an obstacle which is a rigid board one metre square. A wavelength of one metre corresponds to a frequency of about 340 Hz. What will happen is that sound waves of that frequency will be partially reflected, with partial bending round the board. Frequencies around one third of 340 Hz – about 100 Hz – and the wavelength being about 3 m will be almost totally unreflected. For a wavelength of about 30 cm, there will be almost complete reflection, the frequency in this case being about 3 kHz. A consequence of all this is that in any room or studio being used for sound recording the effect of obstacles can be very difficult to predict, as they will act as reflectors for some frequencies and not for others. And in this context obstacles can mean music stands, people, music cases and even instruments themselves!

The bending of sound waves (diffraction)

This can be thought of as a complementary effect to reflection, in that waves which are not reflected will be bent round an obstacle. If the wavelength is much greater than the obstacle size then there will be marked bending round by the waves (there is then little reflection). If the wavelength is small there will be little diffraction but considerable reflection. Figure 1.3(a) shows almost complete diffraction.

DEFINITION
Diffraction is the bending of waves round an obstacle. This bending only occurs to a significant effect if the wavelength is greater than the dimensions of the obstacle (the dimensions as 'seen' by the waves).

Figure 1.4 shows what happens when sound waves pass through an aperture. Possibly contrary to what one might expect, if the aperture is small compared with wavelength there is considerable spreading of the beam of emerging waves; if it is large then there is little spreading. It is often easy to see effects such as this in water when ripples, or even large waves, meet either obstructions or gaps in, for example, the walls of a quay.

There are many easy-to-perform but highly instructive experiments that the reader can carry out for him/herself. For example, try holding something like a clipboard, or any other conveniently shaped rigid sheet, between your ear and a source of sound. It will be found that the high frequencies from the source are reduced at the ear.

Then angle the board so that sounds are reflected off it into the ear and see what range of frequencies are affected. Notice how the character of the sound varies when an aircraft passes behind a house, trees or other objects.

The range of simple observations is almost endless.

It is worth remarking that all waves, and not just sound waves, undergo reflection and diffraction when the conditions are right. With light, for instance, the wavelengths are so small in relation to the size of everyday objects that virtually 100% reflection occurs almost all the time. Diffraction does occur but the effects are generally not obvious.

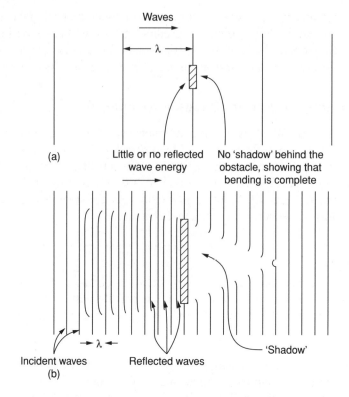

Figure 1.3   Sound waves striking an obstacle

Units used in sound

It will be useful to look at ways of expressing the intensity, or power, of sound waves – but not, at the moment, the thing we call 'loudness'. For various reasons that must come later, in Chapter 2.

Power

This is relatively easy. In all domestic equipment the power required to operate the device is measured in watts (W is the abbreviation) and this will be stated somewhere on it. We are used to 100 W lamps, 3 kW (kilowatt) heaters, and so on.

Since sound waves are a form of energy, we can quite legitimately use watts as units of their power. However, in practice the watt is, in sound wave terms, inconveniently large for most purposes. Or perhaps we should say that the human ear is so sensitive that we need smaller units than watts.

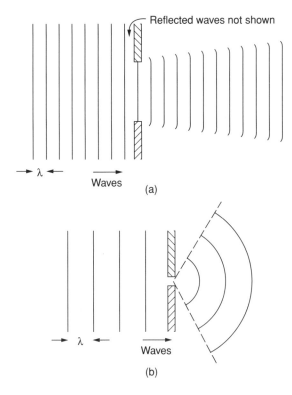

Reflected waves not shown

(a)

(b)

Figure 1.4 Diffraction at an aperture

DEFINITION

Power is the rate at which work is done or at which energy is produced or consumed.

A much more useful unit is the millionth of a watt, or microwatt. This is usually written μW, μ being a Greek letter called 'mu' and generally used to mean 'one millionth'.

At this point, the reader should not become confused, or possibly even alarmed, by the fact that his or her loudspeakers probably have labels quoting much higher powers than the tiny powers, like millionth of a watt, that the ear can handle. Loudspeakers are often rated at many watts, 50 W or more, for even a modest home system. This power refers to the maximum electrical power that can be fed into the loudspeaker without causing either damage or serious distortion. The sound power that comes out will probably be around 1% or 2% of the electrical power that goes into the loudspeaker – and only a very small fraction of that is likely actually to enter the listener's ear!

Intensity

To many people this word would probably be regarded as being synonymous with loudness. However, in scientific terms it has a different meaning. Intensity is defined as the power falling on, emitted by, or passing through a specific area. The units possibly help to make this more clear, being watts/square metre, or microwatts/square metre, as appropriate. We shall write these as W/m^2 or $\mu W/m^2$.

DEFINITION
Intensity is the power falling on, emitted by, or passing through a specific area.

Pressure

In scientific terms this is defined as the force acting on a unit area. Many readers will have seen 'pounds per square inch' (p.s.i.) as a measurement of pressure. (Car tyre pressures are frequently given in p.s.i.: around 28 p.s.i. is a typical value, although the 'bar' is the preferred metric unit.)

Using these units the normal atmospheric pressure at sea level is about 14 p.s.i. Imperial units such as the pound and inch are not recommended for scientific work these days, so we must introduce their metric equivalents. These are, for force, the newton (named after Sir Isaac Newton, 1642–1727) and, for area, the square metre again.

The newton takes a little explaining. It is properly defined as the force which will cause a mass of 1 kilogramme to accelerate at a rate of 1 metre/ second every second. This is not an easy definition to visualize. What is easier to grasp, if less precise, is that an average apple (and one remembers the alleged association between Sir Isaac Newton and apples!) held in the hand exerts a downwards force, because of gravity, of roughly 1 N (N = newton).

Thus, sound wave pressures are normally given as so many newtons/ square metre, abbreviated to N/m^2.

To complicate matters further it has become the official practice to remember the French scientist, Blaise Pascal (1623–1662), by calling the N/m^2 the pascal (Pa for short).

This section on units may seem to have been heavy going. The reader shouldn't worry too much if that is the case, as much of the time in this book we shall be using these units only as a basis for comparison. Familiarity will bring, one hopes, not contempt, but a ready acceptance of them.

DEFINITION
Pressure is the force acting on a unit area.

Sound intensity and the effect of distance

It hardly needs stating that sounds fall off in loudness the further one gets away from the source of the sound. Here we can be a little more precise and state than intensity (see above) decreases with distance at such a rate that for each *doubling* of the distance the intensity (*I*) is *quartered*.

To help understand this, imagine a partially inflated balloon with a square 1 cm × 1 cm drawn on it. Now if the size of the balloon is doubled the square becomes 2 cm × 2 cm – the sides are doubled but the area has increased four times. The intensity (power per unit area) of a sound wave striking the square is thus a quarter of what it was to start with. (There are important provisos about this law which we shall come to in a moment.) This relationship between intensity and distance is expressed by

$$I \propto 1/d^2 \ (\propto \text{ means 'is proportional to')}$$

where *d* is the distance from the source. For example, suppose that at a distance of 1 m from a sound source the intensity *I* is 4 W/m². At 2 m the intensity will be 1 W/m² and at 8 m it will be one sixteenth of a W/m², and so on. This relationship is known as *The Inverse Square Law*.

The important provisos we mentioned above are:

1. The sound source is small compared with the distances involved (theoretically it should be infinitely small). The Inverse Square Law would not apply half a metre from the bell of a tuba! It would, though, give good approximations a few hundred metres from a full brass band.
2. There are no reflecting surfaces in the vicinity.

Consequently, the Inverse Square Law rarely applies very precisely in real life. On the other hand, it can often be used as a valuable approximation.

Decibels

The decibel (dB for short) is a most useful unit, but unfortunately it can cause confusion and dread until it is completely understood. For that reason we shall confine ourselves here to no more about the dB than is absolutely necessary, but more about it is given in Part 2 of the chapter.

The decibel is a unit of comparison – basically of two powers, but it is also applicable to sound pressures and electrical voltages. If two powers, two pressures or two voltages are the same then there is 0 (zero) dB difference between them. In the case of sound it so happens that a change of 3 dB in speech or music is just about detectable to the normal ear. And a change of 10 dB represents to most people a doubling or halving of the loudness of a sound. These, then, are practical applications of the idea of the dB being used to compare sound powers or pressures.

It is particularly convenient at times to adopt standard powers, voltages or pressures to give references against which other powers, etc. may be compared. Such standards are basically quite arbitrary, but are chosen nevertheless to be useful. To give an example, the normal ear at its most sensitive can just detect sound wave pressure variations of 0.00002 Pa (of which more later). This particular pressure is taken as a standard and other sound pressures can be quoted as so many dB above 0.00002 Pa. We are then well on the way towards being able to use decibels as units of loudness.

This is the briefest of summaries about decibels but they will occur many times throughout this book. However, gradual and continued exposure to them will steadily make them more meaningful to the reader. Probably without much effort either.

DEFINITION
The decibel is a unit of *comparison* of two powers, pressures, voltages, etc.

1 Sound waves
Part 2

The velocity of sound waves

This varies slightly with the temperature of the air but not, as is often supposed, with the air pressure, at least not over all normally encountered ranges of temperature and pressure.

At 0°C, the velocity of sound waves in air can be taken as 331 m/s, and this increases by 0.6 m/s for every 1°C rise in temperature.

Thus:

at −10°C the velocity is 325 m/s
at +10°C it is 337 m/s
at +20°C it is 343 m/s.

340 m/s is a reasonable figure to use for normal room temperatures.

In substances other than air, sound waves travel at quite different velocities (Table 1.1).

The high velocity of sound in helium, about three times that in ordinary air, accounts for the high-pitched voices of divers and others whose lungs and vocal tracts contain the gas. (Not an experiment to be tried for more than a very few seconds!)

Table 1.1

Substance	Velocity (m/s)
Water	1480
Glass	5200
Steel	5000–5900 depending on the type of steel
Wood	3000–4000 depending on the wood
Carbon dioxide gas	259 (at 0°C)
Helium gas	965 (at 0°C)

Units

1. The bar. This is another unit of pressure but is most commonly used in meteorology as it is approximately equal to normal atmospheric pressure. It is related to the pascal by:

    1 bar = 100 000 Pa
    or 10 µbar = 1 Pa
    (Television weather charts show air pressures in millibars.)

2. When dealing with sound wave pressures we must of course remember that pressures are *alternating*. Use then has to be made of the concept of a steady pressure which is equivalent to the alternating one. A similar problem is met in the a.c. (alternating current) mains, which in the UK is quoted as having a voltage of 230 V. In fact, the mains voltage goes through a cycle of a maximum of about 340 V, falling to zero and then reaching a maximum of 340 V in the opposite direction. It then goes back to zero and repeats the operation, taking one fiftieth of a second for each cycle (hence the frequency of 50 Hz). This fluctuation can be regarded as equivalent in power to a d.c. (direct current) of 230 V. Engineers use the term *root mean square (r.m.s.)* for this equivalent. The pressures in Pa of sound waves are equivalents in the same way.

Table 1.2   Sound frequencies for different wavelengths

Frequency	Wavelength (metres)
16 Hz	21.25
20 Hz	17.0
50 Hz	6.8
100 Hz	3.4
500 Hz	0.68
1 kHz	0.34*
2 kHz	0.17 (17 cm)
5 kHz	0.068 (6.8 cm)
10 kHz	0.034 (3.4 cm)
16 kHz	0.021 (2.1 cm)

*It is worth noting that this is roughly equal to 1 foot.
The range of frequencies covered in the table (16 Hz to 16 kHz) is approximately the range of frequencies which the normal adult human ear can detect.

Decibels

Correctly expressed, the number of decibels representing a ratio of two powers is

$$dB = 10 \log(\text{power ratio})$$

If we are dealing with pressures or voltages, the expression becomes

$$dB = 20 \log(\text{pressure ratio}) \text{ or } dB = 20 \log(\text{voltage ratio})$$

As an example, suppose that the power amplifier feeding a loudspeaker is delivering 50 W but is then replaced by one delivering 100 W:

the dB increase is 10 log(100/50)
= 10 log(2)
= 10 × 0.3010 (easily found from a scientific calculator)
= 3.010 or approximately 3 dB

It will be remembered from Part 1 of the chapter that this increase of 3 dB is only just about detectable! To get twice the loudness from the loudspeaker, assuming that it will take the increased power without being damaged, the power must be increased by 10 dB. This means that the 50 W amplifier will have to be replaced by a 500 W one:

dB = 10 log(500/50)
= 10 log(10)
= 10 × 1

Questions

Try to answer these as honestly as possible. You *could* cheat by looking at the answers first, but that doesn't really help. The answers are given at the end of the book.

1. What range of sound wave frequencies can the normal adult detect?
    a. 160–1600 Hz b. 16–1600 Hz
    c. 16–16 000 Hz d. 160–16 000 Hz

2. What is the speed of sound waves in air at normal temperatures?
    a. 300 m/s b. 340 m/s
    c. 400 m/s d. 440 m/s

3. A hall being used for music recordings has its gallery supported by square-sectioned pillars. These are 0.5 m by 0.5 m cross-section. What frequency range of sound will be reflected from each pillar?
    a. Below roughly 700 Hz b. Below roughly 350 Hz
    c. Above roughly 700 Hz d. Above roughly 350 Hz

2 Hearing and the nature of sound
Part 1

The mechanism by which sound waves are converted into a perception by the brain is extremely complicated and parts of it are still not fully understood. This book is not the place for even a limited account of the subject. It is, however, a fascinating topic and the reader who wishes to find out more should refer to the list of books for further reading at the end. There are, nevertheless, various aspects of the hearing process which are important in the context of audio technology and these will be outlined here.

The response of the ear

To begin with, we are dealing with a frequency range from about 16 Hz to around 16 kHz. These figures are not precise. At the low end it is not easy to say when sound ceases to have an identifiable pitch and starts to become a sort of flutter. Values of 16–20 Hz are quoted by different authorities. The upper end depends on the individual and also upon the person's age. Twenty kilohertz may be detected by a person in their late teens; some 60-year-olds may have difficulty in hearing anything above 8 or 10 kHz. The effects of exposure to loud noises can also have a very significant effect. Sixteen kilohertz is, though, a fairly good figure to quote as the upper limit of hearing for the average adult.

It is assumed, or at least hoped, that devices such as microphones, tape machines and loudspeakers have what is known as a *flat frequency response*. That is, they respond equally well to all sounds regardless of the frequency. The human ear is not like this. Figure 2.1 shows a graph of one aspect of the normal ear's response to the full range of sound wave frequencies.

The solid curve in Figure 2.1 is called *The Threshold of Hearing*. It shows the minimum sound pressure needed just to create a sensation in the ear. Incidentally, it is a curiosity of the ear that at higher levels the response curve is much flatter.

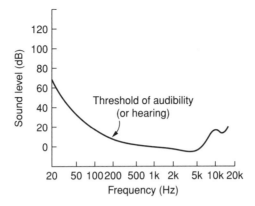

Figure 2.1   The response of the ear to low-level sounds

The horizontal axis represents frequency. The vertical axis is in dB relative to a sound pressure of 0.00002 Pa, which we referred to in Chapter 1. (0.00002 Pa is sometimes written as 20 µPa. A pencil and the back of an envelope can confirm that the two are the same!) This figure represents the sound pressure which can just be detected when the ear is at its most sensitive, which for the normal ear is a frequency of about 3 kHz. An important point to notice about the threshold of the hearing curve is that a very much higher sound pressure is needed to produce a sensation in the ear at low frequencies than at around 3 kHz. In other words, at low sound levels the ear is very insensitive to low frequencies. It is also insensitive to very high frequencies but this is usually less obvious in practice.

The ear's insensitivity to low frequencies can easily be demonstrated by listening to a piece of music with a wide frequency range and turning the volume down to a low setting. It will be found that the music becomes 'toppy'. The high frequency loss which occurs at the same time is less apparent because with most music there is relatively less energy in the high frequencies than in the low ones.

Many domestic audio systems have a 'loudness' control which, when switched in, boosts the low frequencies as the volume is turned down. (Personally I have some doubts as to how useful this really is!)

Loudness. The dB(A)

Loudness is a subjective effect: in other words the loudness of a sound depends upon the individual experiencing it. It cannot therefore be measured. It is possible, however, to make measurements which agree reasonably closely with most people's assessments of loudness.

We implied in Chapter 1 that the decibel might be the basis of a unit of measurement, but a straightforward measurement of sound pressure in decibels is obviously not good enough as it would take no account of the uneven frequency response of the normal ear. If, though, the measurement device incorporates an electrical correction circuit which gives it a frequency response similar to that of the ear, then reasonable approximations to loudness can be made.

Such a circuit was devised many years ago and is incorporated in devices called *Sound Level Meters*. These have a microphone, amplifier, meter and the appropriate circuitry to match approximately the ear's characteristics. Such measurements are quoted with the unit *dB(A)*, meaning decibels with the 'A' circuit incorporated. They are often termed *weighted* measurements. (B and C characteristics have also been devised, but it has been found that the A circuit most nearly matches subjective assessments of loudness.) Table 2.1 lists a few typical sounds with their 'loudnesses' in dB(A).

It may be worth noting that current noise regulations in the UK require ear defenders to be made available in an industrial workplace where the noise levels reach 85 dB(A). Above 90 dB(A) their use is mandatory and also noise levels from machinery should be reduced if at all possible.

DEFINITION
The dB(A) is a reasonably good approximation to most people's assessment of loudness.

Musical pitch

By this we mean the position of a note in the musical scale. Pitch can be a very complicated subject – like so much else in sound! It must be enough here to give the outlines of a few basic facts:

1. Pitch is basically related to frequency, but the perceived pitch of a note can be affected by the loudness.
2. The standard of pitch is taken to be a sound having a frequency of 440 Hz. This is known as 'International A'. It is the note blown, usually by an oboe, as a tuning signal before an orchestral performance. (Oboes have a stable note and also a very penetrating sound which can be heard reasonably clearly by the other members of the orchestra.)
3. Despite what has been said in 1 above it is possible to make up a table relating pitch and frequency. The note designated C is the one called

Table 2.1   Some typical 'loudnesses' in dB(A)

Source	dB(A)
Orchestra, fortissimo at 5 m	100*
Busy street, workshop	95
Orchestra, average at 5 m	80
Domestic TV, at 3 m	70
Speech at 1 m	60–65
Inside quiet car	50
Inside quiet house	30
Inside quiet country church	25

*An orchestra may reach momentary peaks of 120 dB.

'Middle C' on a keyboard instrument. Table 2.2 lists the frequencies for different pitch notes.

4. A musical octave represents a doubling or halving of the frequency. Thus, in Table 2.2, a is an octave below A and its frequency is half of 440 Hz. The octave above middle C has a frequency of approximately 523 Hz.

5. Adjacent semitones (e.g. from one note on a keyboard to the adjacent one above or below, which may mean going from a white note to a black one, or vice versa) have frequencies in the ratio of about 1.06:1. In other words, there is a 6% frequency difference. A note raised by a semitone is indicated by the 'sharp' sign #. A note *lowered* by a semitone is marked '♭'.

The subject of pitch is a complicated one. Table 2.2 shows what is called an *equal tempered* scale, the standard scale for keyboards, and

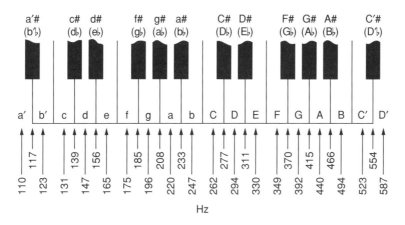

Figure 2.2   Part of a keyboard

on it C# is the same as D♭. However, to a good musical ear these should not really be the same note. Keyboards have compromise tuning!

Musical quality ('timbre')

By timbre (pronounced '*tarmbre*'), we mean the factors which enable the ear to distinguish one instrument from another. A trumpet and a saxophone produce clearly different sounds even if they are playing the same pitch note.

There are two important factors in determining the timbre:

1. Harmonics. These are tones having frequencies which are exact multiples of the base frequency (called the *fundamental*). An instrument playing the note a = 220 Hz will be likely to be producing not only 220 Hz but also many multiples of that frequency, such as 440, 660, 880, 1100, 1320, and so on. Generally, the various harmonics diminish in amplitude as they go up in frequency. Musicians often refer to harmonics as *overtones* but then the first overtone is the second harmonic, and so on.
2. Starting transients. For a brief period at the start of a note – and this may be as little as a hundredth of a second, and rarely more than about one fifth of a second – there are additional frequencies present which are *not* numerically related to the fundamental as the harmonics are. These frequencies die away quickly, hence the name 'transient'. They depend, possibly amongst other things, on the way and rate at which vibrations build up in an instrument. They are a kind of fingerprint.

DEFINITIONS
The *fundamental* frequency in a musical sound is the lowest frequency.
Harmonic frequencies are multiples of the *fundamental frequency*.

For some reason, the ear and brain attach great importance to the starting transients. If they are removed or distorted it is possible for the perceived quality of the instrumental sound to be radically changed, even though only a fraction of a second has been changed at the start of a note lasting several seconds. For example, it can be almost impossible to tell the difference between the sound of a flute and a trumpet, playing the same pitch steady note, if the first few seconds of the sounds are in some way suppressed.

Table 2.2 Frequencies of musical notes

Note	Frequency to nearest Hz
A	**440**
G#	415
G	392
F#	370
F	349
D#	330
D	294
C#	277
C	262*
b	247
a#	233
a	220

*Also known as 'Middle C'.

It is important to realize that the frequencies required to reproduce properly the sound of a musical instrument or the human voice extend, because of the harmonics, well beyond the apparent musical range. For example, the highest note on a piano has a fundamental frequency of a little over 4 kHz. However, the harmonics extend over most of the audio range. For example, taking a note only two octaves above middle C, and using approximate frequencies for simplicity:

$$\text{Middle C} = 262 \text{ Hz}$$
$$\text{one octave above} = 523 \text{ Hz}$$
$$\text{two octaves above} = 1046 \text{ Hz}$$

The tenth harmonic of Middle C (which may or may not in practice be very significant, depending on the instrument) is 10×1046, which is 10 460 Hz. The top notes of a violin have fundamental frequencies around 3 kHz. The fifth harmonic of such notes is about 15 kHz.

A fuller list of musical instrument ranges is given in Part 2 of this chapter.

The brain's perception of sound

Optical illusions, in which the brain superimposes its own interpretation on what the eyes receive as light, are well known. People involved with sound need to be aware that there are aural equivalents.

1. To begin with it is important to realize that the brain's *really* accurate memory for sounds is very short – around 1–2 seconds. For

example, if two good quality but slightly different loudspeakers are to be compared then this must be done by switching directly from one to the other – a process sometimes known as 'making A–B comparisons'. It is impossible to make reliable judgements by listening to one and then the other with a day in between! Comparisons between one good and one mediocre loudspeaker can be made with a time gap, but here we are talking of accurate comparisons between similar quality devices. This is such an important topic that we shall return it to later.

2. It is very easy for the ear/brain system to hear what it thinks it ought to hear! Most professional sound operators know of instances when they have adjusted a control on a piece of equipment and been well satisfied with the result, only to discover later that the control had been switched out of circuit and was therefore doing nothing!

3. Curious effects can sometimes occur in stereo. As an instance, many people listening to stereo recordings of aircraft, in which the image of the aircraft sound goes from one loudspeaker to the other, find that the sound image appears to rise above the level of the loudspeakers and may indeed even seem to be overhead. The explanation is clearly that the brain associates aircraft noise with height and then imagines a height effect even when the real sound sources are at ear level.

4. The ear/brain system has an uncanny ability to discriminate against sounds it regards as unimportant. It can do this because, by having two ears, the brain can locate the direction of the wanted sounds. This effect is sometimes referred to as the *cocktail party effect*. (You can guess why!) Microphones, even when there are two as for stereo, cannot provide this discrimination, or at least to only a slight degree, so that what may seem to be acceptable acoustic conditions when an environment is listened to 'live' may prove to be quite unacceptable, for reasons of background noise or excessive reverberation, when a recording is played back.

5. The remaining item in this short list is known as the Haas Effect. Stated briefly, if similar sounds arrive at a listener's ears then the sounds that arrive first determine the apparent direction of the sound source, even though the later-arriving sound may be as much as 10 dB higher in level. The reader can verify this by feeding a pair of stereo loudspeakers with a mono source and then moving towards one of the loudspeakers so that the sound from it arrives first at the ears. This will cause the mono sound image to move to that loudspeaker. Adjustment of the balance control to raise the level in the other loudspeaker is unlikely to shift the image until there is a very appreciable loudness difference between them.

An accurate comparison of two similar audio devices (such as two loudspeakers) can only be made by switching quickly from one to the other. This is referred to as '*A–B comparison*'.

2 Hearing and the nature of sound

Part 2

Pitch

We said earlier that the frequency ratio between two adjacent semitones was approximately 6%. In scientific (but not necessarily musical) terms, the exact number is

$$^{12}\sqrt{2} \text{ (the twelfth root of 2)}$$

This is the number which, multiplied by itself 12 times, equals 2 and is 1.0594631. A pitch change of one semitone is therefore equivalent to a frequency change of 5.94631%. (Six per cent is a good enough approximation for most purposes!)

The reasoning behind the mathematics is that there are 12 equal semitone 'intervals' in an octave, which itself is a frequency ratio of 2:1. Each step must therefore be $^{12}\sqrt{2}$.

The interested reader with a scientific calculator might try entering 1.0594631 into the calculator's memory and then using Memory Recall 12 times in the following way:

$$MR \times 1.0594631 = 1.122462$$
$$MR \times 1.122462 = ... \text{ and so on, a total of 12 times.}$$

The final result will be 2, or something very close indeed to it, depending on the calculator.

Frequency ranges in music

Table 2.3 gives the approximate ranges of fundamental frequencies of a few musical sounds. The range from lowest to highest notes is often

One octave – 7 white notes and 5 black = 12

Figure 2.3   Part of a keyboard

termed the compass of the instrument or voice. In each case the harmonics will extend the upper frequency, in some cases very considerably.

'False bass'

This is the name given to an interesting effect by which the ear/brain system can somehow seem to insert very low frequencies which ought to be present but which are not reproduced. It may be that the brain recognizes a series of harmonics as being based upon a particular frequency and mentally compensates for this frequency if it is not present.

To take an example, Table 2.3 shows that the lowest pitch notes from a large pipe organ may be as low as 15 Hz. There are very few loudspeakers which can reproduce frequencies that low, and yet our ears are not generally aware of this deficiency.

Table 2.3   Approximate frequency ranges of musical sounds

Source	Frequency (Hz)	
	from	to
Female singer	250	1000
Male singer	100	350
Flute	250	2500
Bassoon	60	600
Alto saxophone	125	650
Trumpet	200	1000
Violin	200	3500
Cello	70	600
Piano	30	4000
Pipe organ (large)	15	8000
Xylophone	700	4000

Questions

1. Decibels could be appropriate to use for comparison in two of the following. Which two?
    a. Two frequencies   b. Two sound pressures
    c. Two loudnesses   d. Two fundamentals
    e. Two powers   f. Two loudspeakers

2. A note has a frequency of 200 Hz. What is the frequency of the note an octave above?
    a. 212 Hz   b. 300 Hz   c. 400 Hz   d. 2000 Hz

3. And what would be the frequency of a note an octave below?
    a. 10 Hz   b. 50 Hz   c. 100 Hz   d. 188 Hz

3 Basic acoustics
Part 1

It would not be far from the truth to say that the secret of a good recording lies in having satisfactory acoustics to start with. But we should make two points: first, it is possible to achieve acceptable recordings in less than ideal acoustic conditions, and perhaps this book will help to show how that can be done; second, what may be good acoustics for one purpose may not be good for another. Again this will, we hope, be made clear. Nevertheless 'good acoustics' are a great help and the aim of this chapter is to explain what is meant by the term and how it may be possible to achieve if not good, then at least adequate, acoustics for particular purposes.

Briefly, we can say that good acoustics enhance, or at least do not impair, recorded sound quality; bad acoustics can reduce the intelligibility of speech, allow distracting background sounds to be present, adversely affect the quality of music, and so on.

We shall look at two aspects of acoustics. These are:

1. Sound isolation – the prevention or reduction of external noise in the recording venue.
2. Internal acoustics – the results of the behaviour of sound waves within this venue.

Sound isolation

It is unrealistic on financial grounds to try to soundproof totally even a professional studio. In practice, broadcasting and recording companies generally adopt criteria which establish how much background noise is acceptable for any particular type of programme material. For example, to aim for an absolutely silent background in a studio to be used for rock music is pointless. On the other hand, it is essential for radio drama studios to have no intrusive noises as these can destroy the illusion: a distant audible police siren cannot help a play set in the eighteenth century! But the siren would probably be utterly inaudible in the rock music studio. There are two categories of background sound which we can consider here:

1. Airborne sound

This means sound which has travelled through the air for the great majority of its journey. The most important methods of preventing its ingress are:

1. Heavy (i.e. massive) walls. More details are given in Part 2 of this chapter, but they can be summarized by saying that a wall made of a single thickness of brick will reduce external sounds by some 45 dB(A) on average. This may sound impressive until it is realized that a moderately busy road could produce a sound level of 80 dB(A) at the outside of the walls. This, reduced by 45 dB, leaves 35 dB(A), which is almost certainly going to be picked up by microphones. In particular, any fluctuations are going to be more distracting than a steady noise. Doubling the mass of the wall has a very modest effect – usually an increase in sound reduction of around 5 dB – so a double thickness brick wall would provide only about 50 dB of sound reduction. Cavities in walls add to the effect, of course.
2. Double or triple glazing of windows. Up to a point, domestic double glazing is not all that effective in acoustic terms as the air gap, which is typically a matter of millimetres thick, does not possess sufficient *compliance* (i.e. 'springiness'). Vibrations in one sheet of glass are easily transmitted to the other. For really good sound insulation the sheets of glass need to be some 200 mm apart, although spacings of 60–80 mm can be fairly useful. For this reason 'secondary double glazing' may be better, provided all gaps are sealed (see below). Notwithstanding what we have just said, domestic double glazing, with its relatively few millimetres of gap, is better than single glazing!
3. Sealing of all air gaps. This is vitally important, as sound waves can pass through extremely small apertures. The fact that domestic double glazing gives an improvement in sound insulation is probably more due to the careful sealing than to the fact that there are two sheets of glass. In professional studios all doors have, or should have, 'magnetic seals', in principle not unlike the seals around refrigerator doors. Double doors, providing a kind of 'sound lock', are common. Further, care has to be taken to seal all places where services such as water, gas and electrical trunking enter the studio.

Reduction of airborne sound in the non-professional environment

Away from a proper studio, what can be done? The short answer is usually not very much. If a room is to be used as a temporary studio the following suggestions may be useful:

1. The obvious, but possibly easy to overlook, first step is to use a room on the quietest side of the building, and if traffic noise varies with the time of day, then quiet times should be chosen if possible.
2. It may sometimes be very worthwhile putting strips of adhesive foam round the edges of all opening windows, unless they are very well-fitting ones. It can happen that an almost dramatic improvement in sound isolation may be achieved for a very modest cost.
3. Noise from adjacent rooms can be a problem. If reducing the noise at source (always the first course of action) cannot be done then temporary sealing round the doors can be tried. Tightly folded sheets of newspaper pushed into the spaces round the door may help – until someone needs to go through the door!
4. Other gaps should be sealed with whatever materials come to hand. Bitumastic sealants can sometimes be very effective for permanent scaling of gaps.

The Haas Effect, mentioned in Chapter 2, must be borne in mind here, as sound may be leaking through a gap but because the first-arrived sound is through, say, a wall, it may not be apparent that the wall is not the major source of trouble.

What are known as flanking paths may be mentioned here. These are routes round, typically, a wall. The wall may be thick, but a possible sound path could be over the top of the wall and through relatively thin ceiling materials.

2. Structure-borne sound

Here the sound vibration travels through the actual fabric of the building, and it is well known that bricks and concrete are good conductors of vibration. (An electric drill with a masonry bit at work on a wall can be heard throughout a house.) Not only that, but water pipes or other rigid material extending through the building can be serious offenders. This kind of noise transmission is the most difficult to try to cure. In fact, cure is often virtually impossible – the problem has to be antici-pated and dealt with at the design stage. In the professional world, the following steps are taken:

1. Studios are sited as far away as possible from likely sources of vibration.
2. They are isolated from the rest of the building by making the studios independent 'boxes' within the building. They stand upon resilient bases of steel springs or some form of synthetic rubber. (Some modern concert halls, such as the Birmingham Symphony Hall, are acoustically isolated in this way.)
3. Care is taken that the studio walls which may be part of a cavity wall system are connected to the other brickwork by flexible tie rods.
4. Flexible seals are used at junctions between studio floors and where they meet, for example, a corridor.
5. Water pipes, etc. have flexible sections.

All this is very expensive, and for obvious reasons can rarely, if ever, be applied as a remedial treatment to an existing building. It is interesting to note that many recording studios built in the last decade or two have been conversions of older buildings, but in rural areas where vibration through the ground, often caused by heavy traffic, is unlikely to be a problem.

As a further precaution against trouble from structure-borne sound, any machinery that has to be near a studio, for example fan motors for ventilation and conditioning, are fitted with *anti-vibration mounts* (AVMs) so that the motor is, as far as possible, isolated from the floor on which it stands.

Reduction of structure-borne sound in the non-professional environment?

It is clear that very little can be done about structure-borne sound. Noise from machinery can of course be stopped by switching off the machine – but this may not always be practicable. Microphones can often pick

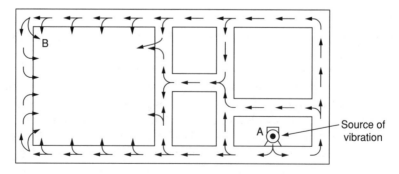

Figure 3.1   Structure-borne sound

up severe rumble by sound transmission through the floor and via a table to the microphone stand. It can happen that this is very disturbing at the playback of a recording when it may not have been obvious at the time of making the recording. Suitable resilient supports for the microphone help, but putting the first rubber mat that comes to hand under the microphone stand may not be the answer. The best kind of sound-insulating support is one where the microphone and its stand 'float' on the mat, so that the stand sinks slightly into it but does not compress it fully. A little experimenting with different materials is often worthwhile.

Internal acoustics

This is basically a matter of sound reflections from walls, floor, ceiling and other surfaces, together with, as we shall see, absorption of sound at such surfaces. We shall consider two of the aspects which, in the context of this book, are the most important. These are:

1. Standing waves

When there are two parallel facing surfaces, such as opposite walls, it is possible for there to be a number of modes of vibration, called *standing waves*, as shown in Figure 3.2. If the wavelength of a sound created between these surfaces is such that multiples of the half-wavelength correspond exactly to the distance between these surfaces, then it is possible for an acoustic resonance to develop.

There are many instances of these resonances. Organ pipes are good examples. Also, the note which is produced when one blows across the top of an empty bottle is the result of a resonance, although the mechanism

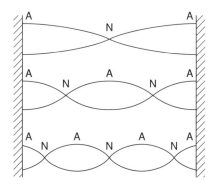

Figure 3.2 Standing waves

by which it is produced is very different from that which can occur between two walls.

The term *standing waves* for the vibration patterns shown in Figure 3.2 is somewhat inaccurate, because the only things that are stationary are the patterns of minima and maxima. These are called *nodes* and *antinodes* and are marked N and A in the diagram. Figure 3.2 shows the *pressure* variations (there are other ways of representing standing waves) so that an antinode is a point where there is the greatest pressure variation and a node is where there is zero (or at least minimum) pressure variation. (There are other ways besides the sound pressure that can be used to represent standing waves. For instance, the *air particle displacement* can sometimes be useful.)

It is possible to detect standing waves in most rooms if a steady pure tone (i.e. one with no harmonics) is played. Walking about in the room, especially if one ear is covered tightly with the hand, shows variations in the loudness as one moves through the nodes and antinodes. This effect is best noticed when the wavelengths are in the region from about 0.1 to 1 m – a frequency range from around 3 kHz down to roughly 300 Hz.

Standing waves can also be observed in water when the conditions are right. Circular wave patterns in a vibrating cup of liquid are good examples of a rather special set of standing waves.

In acoustic terms, standing waves are a bad thing as there will generally be different nodes and antinodes for different frequencies. Thus, a microphone placed between two parallel reflecting surfaces is likely to give a distorted version of what it should be picking up because it will be at nodes (minimum loudness) for some frequencies and antinodes (maximum loudness) for other frequencies.

DEFINITIONS
Standing wave – in the case of sound, a pattern of maximum and minimum pressures resulting from reflections of the sound.
Node – a minimum in a standing wave.
Antinode – a maximum in a standing wave.

Standing waves are reduced by having non-parallel surfaces and/or irregularities in walls, ceilings, etc. Also, the presence of sound absorptive materials (see later) has a beneficial effect, up to a point. Irregularities, to be effective, need to be large – a figure of one seventh of the longest wavelength is sometimes quoted for this – so taking the lowest speech frequencies as being about 300 Hz, with a wavelength of roughly 1 m, the surface irregularities for speech should be about 15 cm. Ingenuity

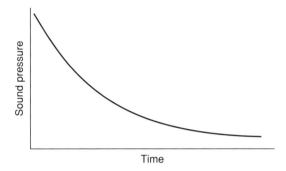

Figure 3.3 Reverberation

can help in the non-professional environment. Cupboard doors ajar and the tops of trestle tables set at an angle against a wall are possibilities.

2. Reverberation time

This is the second important aspect of internal acoustics that we shall deal with. In any room, sound energy dies away rather in the manner shown in Figure 3.3. The curve is somewhat idealized as the decay is always much more irregular than this. This is a result of multiple reflections from all the surfaces, the sound losing energy at each reflection.

Reverberation time, which we shall denote by RT, is defined as the time it takes for the sound to decay through 60 dB. This may be made a little clearer by a look at Figure 3.4. Here the decay of a sound is shown with a decibel scale for the vertical axis – in Figure 3.3 the vertical axis was pressure. The most obvious change is that the sound decay now appears as a straight line (in practice never quite as straight as this!).

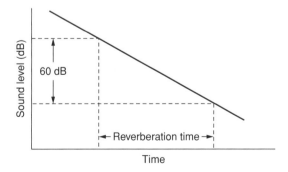

Figure 3.4 Reverberation time

More details and methods of calculation are given in Part 2 of this chapter. At the moment, though, it is enough to say that 60 dB represents roughly the difference between a moderately loud sound and virtual inaudibility.

Two important factors which affect the reverberation time are:

1. The volume of the room or studio.
2. The amount of sound-absorbing material present.

Taking these in turn, it is not difficult to see that the bigger the room the longer it takes sound to reach the various surfaces at which energy will be absorbed. If, for instance, 2 dB of energy is lost at each reflection, then a loss of 60 dB will take place after 30 reflections. The time for 30 reflections to occur will be longer in a large room than a small one.

Sound-absorbing materials present in a room will clearly help the reverberant sound to die away more quickly. We might say briefly at this stage that all materials absorb sound to some extent, although materials like glass and hard woods have only a very small absorption. Porous and fibrous materials are likely to be good absorbers. Hence carpets and curtains and soft furnishings are helpful, although not likely in themselves to provide all the absorption which might be needed. More details are given in Part 2 of the chapter.

DEFINITION

Reverberation Time (RT, or sometimes, T_{60}) is the time taken for the reverberant sound in a room, studio, etc. to decay through 60 dB.

Reverberation time is very important in any sound recording work. For a recording to sound convincing, the RT should be appropriate to the kind of material being recorded. Examples are:

1. Speech. Too long a value of RT gives the speech a distant, 'echoey', effect. Too short and the speech quality is apt to sound dry and lifeless, besides which too short an RT is unpleasant for the speaker. Generally, an RT of around 0.3–0.4 of a second is considered satisfactory.
2. Orchestral and choral music. Around 2 seconds is preferred, although there can be some latitude in this. Most traditional concert halls have RTs of this order. Too short an RT is apt to result in an inadequate blend of individual notes. However, much chamber music and

baroque music written for small ensembles may often benefit from a shorter time. It is worth noting that, with very large halls (the Royal Albert Hall in London is a good example), speech on the platform can seem to be in a very 'dead' (i.e. short RT) acoustic because the low sound levels are not able to reach the distant reflecting surfaces. With an orchestra, though, the Royal Albert Hall's full RT of over 2 seconds is apparent!

3. Pop and rock music. A short RT is needed here, at least in a studio, as it is normal practice to multi-mic the band – that is, each instrument has a microphone. A long RT implies that there is considerable reflection of sound in the studio and as a consequence each microphone will be likely to pick up quite a lot of the sound from other instruments, which is undesirable. Studios for this sort of music have RTs of, typically, around 0.5 second.

4. Drama (sound only). This presents a problem as outdoor conditions may need to be simulated and this requires a short RT. At other times it may be necessary to represent something like a church interior, and this calls for a long RT. It might seem that in the latter case the addition of artificial reverberation could solve the problem, but this may not be satisfactory as an actor often needs a reasonably good simulation of the acoustic conditions in order to help with the delivery of his/her lines. Some typical RTs are given in Part 2 of the chapter.

An interesting, and sometimes very important, point about realism in reverberation is in the Initial Time Delay (ITD). In any place where there is reverberation (and this means pretty well everywhere!) a listener will hear first the direct sound from the source and then there will be a small time interval before the reverberation arrives. This interval is the ITD. The ear and brain take note of an ITD – even if it is too short to be consciously perceived (less than about 30 ms) – and use it to help assess the size of the environment.

In a small room the ITD will be short, perhaps only a very few milliseconds; in a large hall it will be perhaps a few tens of milliseconds.

This means that, in sound, to simulate accurately a particular environment not only does the RT have to be correct, but the ITD must also be appropriate. Artificial reverberation devices usually have a time delay control which in effect varies the ITD.

The next question is, what can be done to give a conventional room or hall the right sort of RT for a particular recording? Let us say straight away that the achievement of the ideal acoustic conditions is a very expensive task (some might say an almost impossible one, even with plenty of money available!) and in what might be termed ad hoc circumstances the range of possibilities is limited. To a large extent good microphone technique can be a great help, and this will be covered in later chapters. The following suggestions may be helpful:

1. Speech. The usual fault with simple equipment is allowing too much reverberation to be picked up. This may be a matter of microphone positioning as much as anything, but that will be gone into later. The non-professional recordist often has little choice of 'studios' and an office or a sitting room may be all that is available. The average sitting room has an RT of around 0.5 second – rather on the long side for speech recording. Offices may be very reverberant if they are large and have little in the way of soft furnishings; they can be quite dead, acoustically, if they are of the plush, executive variety.

    Any or all of the following can help to increase the sound absorption, if that is what is needed: curtains drawn across windows, rugs, blankets, etc. laid on uncarpeted floors, plenty of coats on all hooks, tables covered with thick cloths, rugs or blankets.

2. Music. Choral or orchestral recordings are probably going to be made in sizeable rooms – halls, churches and so on. The RTs may be theoretically too long, especially in the absence of an audience, but it is generally accepted that, for music (classical music anyway), too long an RT is preferable to one that is too short. The opposite tends to be true for speech.

    It is worth noting that it is possible to have a small room with many hard surfaces, yet which has a longer RT than a much larger room with many absorbent surfaces. However, the pattern of reflections in the small room will give it a 'small-room sound'.

A final point here – and this should be seen as being applicable to all recording work – it is very desirable indeed that test recordings are made and listened to critically before proceeding with the proper recording. The ability of the ears to discriminate against unwanted sounds has already been referred to. A microphone doesn't have this ability!

3 Basic acoustics
Part 2

Sound isolation

It was stated in the first part of this chapter that the heavier a wall is the better it is at preventing airborne sound from travelling through it. Strictly we should talk not about the weight, but the mass/unit area, using units such as kg/m² – in other words the mass of a square metre of the wall's surface.

The graph in Figure 3.5, known as *The Mass Law*, shows the approximate relationship between reduction in sound transmission and mass/unit area of the wall. This is only an approximate relationship as there are various lightweight structures consisting of two or more leaves of material which combine to give far better sound isolation than the combined weight (mass) of the leaves would suggest.

Sound absorption

As a general rule, any porous material can provide some sound absorption and it follows that, for example, gloss painting an otherwise porous

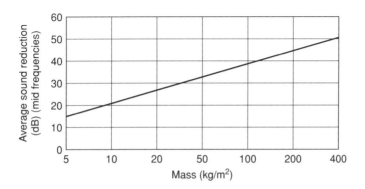

Figure 3.5 The Mass Law

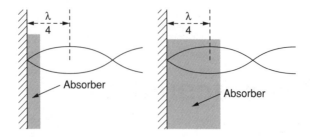

Figure 3.6   The operation of a porous absorber

The curves show the *displacement* of the air particles and not pressure. This is done to make the point that there is maximum particle movement at a quarter of a wavelength (λ/4) from the wall. Greatest frictional effects will therefore occur when this maximum movement is inside an absorber.

surface will immediately reduce its absorptive abilities. Also, a single material will be an effective absorber only if it is thicker than about one quarter of a wavelength (see Figure 3.6). Thus, to absorb sounds of frequencies 300 Hz and above, when the longest wavelength is about 1 m, the absorbent material should be about 25 cm thick. In practice, there is some latitude in this and, for instance, curtains draped in 10-cm-deep folds may be found to be helpful. The same fabric, stretched, will be much less effective. It should be added that there are various types of *wide-band porous absorber*, consisting of more complex structures but capable of absorbing sound over a wide range of frequencies and at the same time not being unacceptably deep. The interested reader can find out more from some of the books listed under 'Further reading'.

The effectiveness of a porous absorber can be increased if it is spaced from the wall by an amount equal to its thickness. Figure 3.7 illustrates this.

A useful indication of the sound-absorbent properties of a material is the *sound absorption coefficient*, often denoted by the Greek letter α (alpha). This is a number less than 1 and is a measure of the amount

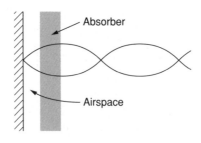

Figure 3.7   The effect of an airspace

Table 3.1 Some absorption coefficients

Material	α (average)
Plain brickwork	0.04
Rough concrete	0.05
13 mm building board	0.2
Wood (typical)	0.1
Curtains, velour, draped	0.6
Curtains, sailcloth, draped	0.16
Wilton carpet with underfelt	0.6

of the incident sound which is absorbed by the material. Thus, if $\alpha = 0$ there is no absorption and all incident sound is reflected (or diffracted), depending on the size of the material. Similarly, if ($\alpha = 1$ there is total absorption. Sometimes α is quoted as a percentage, e.g. $\alpha = 72\%$ is equivalent to $\alpha = 0.72$.

Table 3.1 gives a few typical values of α. Note that with most materials α depends on the frequency of the sound, usually increasing with frequency. In Table 3.1, the absorption coefficient is given for a frequency in the middle of the range – about 400 Hz is often used as a convenient mid-range frequency.

It is a common misconception that polystyrene ceiling tiles are useful acoustic devices. This is not the case. They are too light in weight to provide any sound isolation and they are generally not porous enough or thick enough to provide significant absorption.

Unit of sound absorption

The basic unit of sound absorption is one square metre of 100% absorber. The name given to this unit is the *sabine*, after W. C. Sabine, the American physicist who did pioneering work in acoustics at the end of the nineteenth century. It was he who first introduced the concept of reverberation time and went on to realize that it depended on the volume of the room and the absorption in it. He did this without the aid of any modern electronics, training himself and his assistants to time the decay of a sound by ear. This was before decibels were thought of and Sabine used the time for a sound to decay to one millionth of its original intensity. (A ratio of a million to one corresponds to 60 dB!) For a sound source he had an organ pipe driven by air at a constant pressure.

$$1 \text{ sabine} = 1 \text{ m}^2 \text{ of material for which } \alpha \text{ is } 1.0$$

The number of sabines provided by a surface can be found by multiplying the area by the absorption coefficient. Thus, a Wilton carpet

($\alpha = 0.6$) having an area 4 m by 5 m will have an absorption at mid-frequencies of

$$4 \times 5 \times 0.6 = 12 \text{ sabines}$$

A person seated is equivalent to about 0.5 sabine, a fact which designers of auditoria need to bear in mind.

Reverberation time

A means of calculating RT is given by *Sabine's Formula*. The version given here is only approximate but is useful nevertheless so long as RTs are not very short.

RT (seconds) = 0. 16 × Volume in m³/Total number of sabines

The total number of sabines, n, is found from multiplying the area A of each type of surface by its absorption coefficient and adding the whole lot together. In mathematical terms:

$$n = \Sigma(A_1\alpha_1 + A_2\alpha_2 +...)$$

Non-mathematicians need not be frightened by this! The symbol Σ means 'add together the following', while '$A_1\alpha_1$' means 'area number 1 multiplied by its absorption coefficient', and so on.

It is usual these days for acoustics specialists to use microprocessor-controlled equipment to measure RT rather than calculate it. Methods are beyond the scope of this book, but specialist works on acoustics can be referred to for details.

Table 3.2 Some typical values of RT

Environment	RT
Large cathedral	5 s (or can be much more. I found a cathedral in Esztergom in northern Hungary where the RT was about 15 s, when the place was virtually empty)
Concert hall	2 s
Theatre	1 s
Average sitting room	0.5 s
Open air	almost zero for quiet sounds but can be several seconds for loud noises such as thunder and gunfire

Questions

1. In a standing wave a node is
    a. A maximum of vibration   b. A minimum of vibration

2. Reverberation time is the time taken for reverberant sound to decay through
    a. 6 dB   b. 30 dB   c. 60 dB   d. 80 dB

3. What would be a suitable reverberation time for a talks studio?
    a. 0.1 s   b. 0.5 s   c. 1.5 s   d. 2.0 s

4. The unit of sound absorption is the sabine. It is equivalent to
    a. 1 square metre of 100% absorber
    b. 1 square centimetre of 100% absorber
    c. 1 square metre of 0% absorber
    d. 1 square centimetre of 0% absorber

4 Microphones
Part 1

Microphones are amongst the most important tools in any sound recording work and an understanding of their properties is vital. Fuller technical details can be found in books listed under 'Further reading' – here we must confine ourselves to an outline.

There are two principal characteristics of any microphone. One is the *transducer system* – in other words, the means by which sound waves are converted into electrical signals. The other is the *polar response* – the way in which the microphone responds to sounds arriving from different directions. We will look at these two things in turn, but first we will look briefly at the main components of a microphone. These are:

1. The *diaphragm*. This is a thin circular sheet of metal or plastic which is caused to move by the varying pressures of an incident sound wave. It is invariably placed behind a protective metal grille or a piece of metallic gauze. The diaphragm either forms part of, or is mechanically linked to, the transducer.
2. The *transducer*. This can be one of a number of forms which are outlined below.
3. The *microphone casing*. Besides protecting the diaphragm and transducer, the nature of the casing affects the polar response of the microphones, as explained in Part 2 of this chapter.

Microphone transducers

The following is a brief description of the most important types.

1. Moving coil

Figure 4.1 shows a simplified section through a transducer of this type. The diaphragm is slightly domed to give it extra rigidity and fixed to it is a coil of very thin wire, often of aluminium because of this metal's

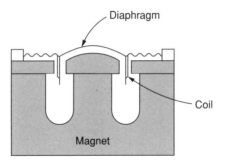

Figure 4.1 Moving coil microphone

lightness. The outside area of the domed section is corrugated to allow the diaphragm to vibrate. As it does so, the movements of the coil in the magnetic field cause a voltage to be generated in it. This voltage is extremely small – of the order of 1 millivolt (one thousandth of a volt) or less – and is a replica of the diaphragm movements and hence of the sound waves which are striking it.

Moving coil microphones are sometimes called 'dynamic' microphones. This doesn't seem to me to be a very good term, as 'dynamic' simply means a system in which things are constantly changing – as opposed to 'static'. Any microphone, therefore, is in a sense *dynamic*. It's far better to call these microphones 'moving coil', which says what they are!

2. Ribbon microphones

Figure 4.2 shows the idea.

Here the very thin corrugated ribbon is itself the diaphragm. Its movement to and fro causes a voltage to be induced in it, but because the

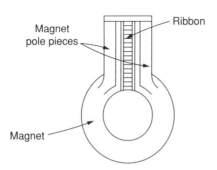

Figure 4.2 Ribbon microphone

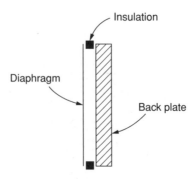

Figure 4.3 Basic electrostatic microphone

total length of the ribbon is relatively small, being rarely more than a couple of centimetres, the induced voltage is also very small, generally much less than that produced in the coil of a moving coil microphone.

3. Electrostatic ('condenser') microphones

The diaphragm and backplate form a capacitor. As the diaphragm vibrates the capacitance varies and, as is explained a little more fully in Part 2, a corresponding voltage is produced. *Electret* microphones are a form of electrostatic microphone and are mentioned further in Part 2.

Other types of transducer exist but they are not of great importance today. For the sake of completeness though, we might mention the *carbon granule microphone*. A layer of small grains of carbon is held between the diaphragm and a rigid plate. Vibrations of the diaphragm cause the pressure on the granules to change and this varies their electrical resistance. A current (produced by, say, a battery) flowing through the granules is thus caused to vary also. For many years, this type of transducer was used in telephones and it was widely used in the very early days of broadcasting. Its drawbacks are that it generates a very detectable hiss in its output, apparently caused by minute electrical arcs between the granules, it is less reliable than more modern types and it is not easy to have different polar diagrams – a point which will be better appreciated when that topic is dealt with a little later. Also, the granules tended to pack together so the whole thing had to be tapped firmly from time to time. The carbon microphone's advantages were that it gave a high output and it was cheap. It is very rarely encountered now.

Another type of microphone which is obsolete in the broadcasting/ recording world but might be found occasionally in second-hand shops is the *crystal microphone*. Carefully chosen slices of crystals of certain materials show what is known as the *piezo-electric* effect. In other words,

Table 4.1 Comparison of moving coil, ribbon and electrostatic microphones

Moving coil

For:	(a)	Usually very reliable and robust (although all microphones should be treated with as much care as possible).
	(b)	Can be used on the end of considerable lengths of cable without the need for amplifiers close to the microphone (unlike electrostatic microphones which, for reasons given in Part 2, have to have some form of amplifier near to the microphone).
Against:	(a)	Tend to be expensive as skilled labour is needed in their assembly.
	(b)	The quality of their output, although as a rule good, is not likely to match that of a high-grade electrostatic microphone.

Ribbon

For:	(a)	The quality can be very good indeed because of the extreme lightness of the ribbon.
Against:	(a)	Rather fragile.
	(b)	Cannot normally be used out of doors as the slightest wind on the ribbon causes serious rumble noises in the output.
	(c)	Usually expensive.

Electrostatic

For:	(a)	Excellent quality is possible.
	(b)	While the very best microphones of this type are costly, reasonable quality microphones can be quite cheap.
Against:	(a)	For reasons explained in Part 2 it is necessary to have an amplifier close to the diaphragm assembly, either within the body of the microphone or not more than about a metre away. Generally, though, when one buys an electrostatic microphone one gets the complete unit.
	(b)	Apt to be affected by moisture – not necessarily permanently but if, for example, one is taken from a cold environment into a warm room there is a likelihood of condensation causing loud crackling or 'frying' noises in the output until the microphone has dried out, which, even in a warm room, might take some time. Placing the microphone on a warm (but not too hot!) radiator speeds up the drying-out process.

when the piece of crystal is deformed a small voltage is generated. Quartz is one such material and the ubiquitous quartz watch makes use of this electro-mechanical property to give very accurate time keeping. The crystal microphone suffers from a number of disadvantages, one being that crystals appear to vary in their acoustic properties, so that while quartz is good in watches it is less satisfactory than other types of transducer in microphones.

Polar responses

It might be imagined that a microphone ought to respond equally well to sounds arriving from all directions. In some cases this is what is wanted, but lack of sensitivity to certain angles of incidence can be invaluable in discriminating against unwanted noises. The best way of illustrating the directional characteristics of a microphone is by means of a *polar diagram*, such as that shown in Figure 4.4. The microphone itself is imagined to be at the centre of the diagram and the distance from there to the heavy line at any angle is proportional to the sensitivity of the microphone to sounds arriving at that angle. The scale from the centre outwards may be in voltage or, more usefully, in decibels. The line marked 0° is taken to represent the direction in which the microphone is facing, so that 90° and 270° represent the sides and 180° the rear.

There are several basic polar diagrams which are (a) reasonably easily achieved by manufacturers and are (b) found to be useful for different applications. Some of these are listed below, but a fuller explanation is given in Part 2.

1. Omnidirectional

Figure 4.4 shows the polar diagram for a microphone which responds equally well to sounds from all directions – hence the name *omnidirectional*. Note the little sketch of the microphone with the significant directions.

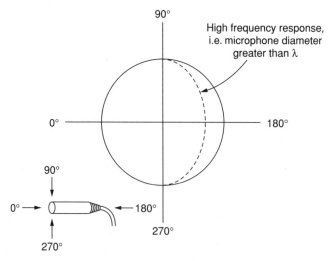

Figure 4.4   Polar diagram for an omnidirectional microphone

In practice, a suitably designed microphone of this type, unless it is very small indeed, can be omnidirectional only up to a certain frequency. This is because, to respond to sound waves arriving from behind (180°), the sound wavelengths must be appreciably longer than the dimensions of the microphone if they are going to diffract round it. The dotted line in Figure 4.4 shows, in a much simplified way, the response that can be expected at the higher audio frequencies from a microphone having a diameter of 2 cm.

(An important point is that polar diagrams have to be drawn in two dimensions, but they actually refer to a three-dimensional concept. Thus, a true polar diagram for an omnidirectional microphone is a sphere and not a circle. This should be remembered with all other polar diagrams.)

2. Figure-of-eight

As will be seen from Figure 4.5, the name is self-explanatory.

This pattern is achieved by allowing sound waves to reach *both* sides of the diaphragm. Ribbon microphones lend themselves particularly well to being figure-of-eight, but electrostatic microphones of the more expensive type can also be made to have this response. It can be a very useful pattern as unwanted sounds from the sides are discriminated against.

Face-to-face interviews in poor (e.g. excessively reverberant) conditions are an obvious application. However, figure-of-eight microphones tend to be rather prone to producing severe rumble noises in their output when they are handled and also, being susceptible to the effects of wind, they are not suitable for outdoor work. Consequently, they are often of little use except in a good studio environment. Furthermore, the high

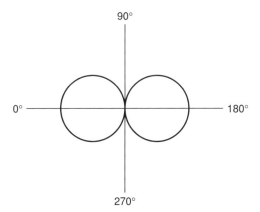

Figure 4.5 Figure-of-eight polar diagram
(A true representation would be two touching spheres)

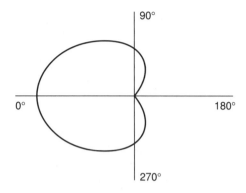

Figure 4.6   Cardioid response

cost of almost all figure-of-eight microphones tends to limit their use in the non-professional world.

3. Cardioid

Figure 4.6 shows the heart-shaped pattern which explains the name. (It's taken from a Greek word meaning 'heart' – hence words like 'cardiogram'.)

This is probably the most useful of all the responses listed here. It should be pointed out, though, that the response at 180° is, in practice, never zero. The reduction in sensitivity compared with sounds from the front (0°) is unlikely, even in a very expensive professional microphone, to be more than about 25–30 dB and at some frequencies (usually the low ones) will almost certainly be less. Nevertheless, a reduction of even this is enough to be very useful. Remember that, subjectively, a reduction of 20 dB will make a sound seem about a quarter as loud.

The 'dead' (rear) side can often be directed towards unwanted noises and the response at the sides (90° and 270°) of the microphone is lower than the 0° response by about 6 dB, which is helpful at times.

4. Hypercardioid

This is a diagram which is intermediate between figure-of-eight and cardioid (Figure 4.7). Its most obvious characteristic is that there are two 'dead' angles, each at about 45° off the rear axis. The term 'super-cardioid' is sometimes used to imply a similar pattern.

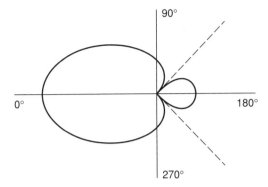

Figure 4.7 Hypercardioid pattern

5. 'Gun' patterns

The name is derived from the shape of the microphone required to produce this pattern, being usually an extended tube with a diameter of 2–3 cm. There are two general versions characterized by the length of the tube: 'short' guns, about 25 cm long, and a longer type with a tube length of around 50 cm. The short tube microphones have a polar response which is not unlike a hypercardioid – and indeed are often described as being this. The longer tube microphones have a more directional response. Typical curves are shown in Figure 4.8.

Note that each type becomes more directional as the frequency increases, and both are relatively non-directional at low frequencies. Microphones of this kind are very widely used, especially for outdoor drama and news-gathering operations, when their marked ability to reduce the effects of sounds which do not arrive from the front of the microphone

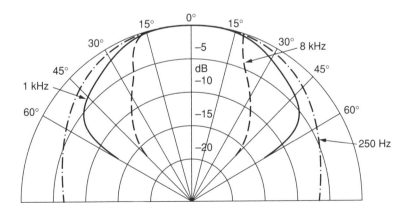

Figure 4.8 Typical 'gun' microphone responses

allows them to be used much further from the action than is possible with other types.

Gun microphones tend to be very expensive indeed, but can often be hired at quite reasonable rates for the day or the week. One very important practical point to note is that these microphones tend to lose their directional properties in small rooms – but one might wonder why anyone would want to use such a device for sound pick-up in a small room!

'Boundary' microphones

There is a category of microphones which has achieved significance in the last decade or two, consisting of a conventional microphone unit (usually electrostatic and inherently omnidirectional) mounted very close to the centre of a metal plate or in the surface of a block of wood. The term 'boundary' is only one description. They are sometimes referred to by the trade mark name of 'Pressure Zone' microphones. Such microphones are best used when placed on a large flat surface such as a table or even the floor, when their polar response becomes, almost inevitably, hemi-spherical. They are fairly immune to vibration effects and can thus be useful for round-the-table conferences or other meetings, when the microphone is placed, naturally, in the centre of the table. Adequate sound pick-up, perhaps not for serious reproduction but quite good enough for subsequent preparation of a transcript, can be obtained with the microphone on the floor in the centre of a room, with perhaps as many as 20 or 30 people present and all likely to contribute to the discussion.

Personal microphones

The little microphones which are often clipped to a presenter's clothing come into this category. They are usually omnidirectional electrostatic devices and have two advantages: (1) they are very inconspicuous and (2) if the person moves about then the microphone moves with them. They are sometimes called 'lapel microphones' or 'tie-clip microphones'. Another term, really quite incorrect, is 'Lavalier microphones'. This term was used for the first personal microphones which were on a kind of lanyard worn round the neck. It is said that the word came about because a certain Madame Lavalier in the court of a French King used to have a pendant made of very large jewels!

Radio microphones ('wireless microphones')

Both terms are slightly inaccurate, although it is not easy to think of convenient and more suitable ones! The Americans use the term 'wireless microphone' and one sees what they mean – there are no cables, but there are wires inside the device!

To begin with, the actual microphone is perfectly conventional. In principle, any type could do. The 'radio' refers to the link between the microphone and the next part of the chain of equipment. This link is usually an f.m. (frequency modulated – see Part 2) radio system, consisting of a small, low-power transmitter at the microphone end and a matching receiver at the other end. The transmitter is battery operated and may be built into the microphone body – in which case the aerial is commonly a short rod or piece of flexible wire sticking out of the base of the microphone – or it may be a separate unit to go into a pocket or be otherwise hidden in the clothing. The latter system is the preferred one if, for visual reasons, the microphone and its transmitter have to be as inconspicuous as possible. The microphone itself is then likely to be of the 'personal' type, which can be hidden if necessary.

The receiver generally has a short aerial fitted to the case and it may be mains powered. The range of the average radio microphone is around 30 m – greater than this in good conditions, less in others. Care has to be exercised when more than one transmitter/receiver pair is used at the same site, as each needs to use a different radio frequency to avoid interference. Also, it is not unknown for radio microphones being used by a quite different organization in the near locality to cause mutual interference.

Good radio microphones are expensive and unless they are going to be used regularly – say more than once or twice a week throughout the year – it is best to hire them.

Questions

1. Which of the following statements are correct about moving coil microphones?
   a. They need to have an amplifier close to them
   b. They are generally robust and reliable
   c. The quality of their output, while good, is likely to be inferior to some other types

2. Which of the following statements are correct about ribbon microphones?
   a. They are usually fragile
   b. Very good quality is possible because of the lightness of the ribbon
   c. They are cheap

3. Which statement best describes a figure-of-eight polar pattern?
   a. It is like a figure 8 on its side
   b. It is like a figure 8 upright
   c. It is like two spheres in contact

4. A good cardioid microphone's sensitivity to sound from the rear is lower than the front sensitivity at best by about
   a. 10–12 dB   b. 25–30 dB   c. 50–60 dB   d. more than 80 dB

5. A hypercardioid microphone has two 'dead' angles. These are approximately
   a. 45° on each side of the front
   b. 90° on each side of the front
   c. 45° on each side of the rear
   d. 270° on each side of the rear

4 Microphones

Part 2

Electrostatic microphones

Figure 4.9 shows a typical, if simplified, circuit for an electrostatic microphone.

The capacitor part, known as the *capsule*, consists of the conductive diaphragm and the backplate. A d.c. supply, which is usually in the region of 50–100 V, provides a polarizing voltage on the capacitor. The charge on a capacitor is given by

$$Q = CV$$

where Q is the charge (in coulombs), C is the capacitance in farads and V is the voltage across the capacitor. The resistance R in the diagram is of extremely high value, a few hundred megohms, so that when the microphone is 'switched on' it takes an appreciable time (compared with the duration of sound wave cycles) for the capacitor to become fully charged. In short, the charge Q is, as it were, locked in. Consequently, if C varies, as it will do when the diaphragm vibrates, because Q is constant (it is determined by the area and spacing of the diaphragm and backplate), it follows that V, the voltage, must vary. It is this variation in voltage which constitutes the microphone's output.

The electrical impedance of the arrangement is very high and this means that the cable from the microphone to any other equipment will

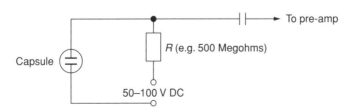

Figure 4.9 Basic circuit of an electrostatic microphone

be seriously liable to pick up external interference – a 50 Hz hum from the mains being a common problem. For that reason a small amplifier – the *pre-amp* as it is often termed – has to be installed close to the capsule. There is now a new problem – providing power for this amplifier. Many microphones incorporate small batteries, but there are other ways of supplying power and these will be dealt with later in this section ('Phantom power').

Electret microphones, now very common, use special materials which carry a permanent electric charge on either the backplate or the diaphragm. Hence the term 'electret', by analogy with 'magnet'. The use of these materials removes the need for a d.c. polarizing voltage across the capsule. A pre-amp is still needed, though, and this means a power supply of some sort, albeit no more than a small battery. Costly electret microphones are capable of extremely high quality, but it is also possible to buy quite inexpensive ones with respectable performances.

There is a further kind of electrostatic microphone – the *r.f. electrostatic microphone* (r.f. = radio frequency). The mode of working is too complicated to go into here and it must be enough to say that the capsule is used to vary the tuning of a radio frequency circuit – hence the name. They are expensive devices but they have the great advantage of being largely unaffected by humidity. Some types of gun microphone are of the r.f. variety and these are much used by broadcasters for outdoor work, where humidity might otherwise be a problem.

Production of the different polar responses

1. Omnidirectional

Perhaps paradoxically, sound waves must be allowed to reach only the front of the diaphragm. Diffraction (see Chapter 1) is relied upon to let sounds arriving off the main axis bend round and strike the diaphragm. At high frequencies, as we have said, this effect may not occur, or only partially. In some situations the failure to have a perfectly omnidirectional response at the higher frequencies may not matter.

Omnidirectional microphones are also termed *pressure operated microphones*, as it is the sound wave pressure and not some derivative of it that actuates the diaphragm. This point may be clearer after reading the next two sections on figure-of-eight and cardioid microphones.

2. Figure-of-eight

Both sides of the diaphragm are exposed to sound and what causes it to move is a force based on the *pressure gradient* – in other words the difference in acoustic pressures on the two sides. This means that the 0° and 180° angles of incidence give equal microphone outputs. Sounds arriving at angles of 90° or 270° will produce the same pressures on both sides of the diaphragm and thus will have no resultant force acting on it and it will not move: hence the lack of output for these angles of incidence. In mathematical terms the equation for a figure-of-eight pattern is given by

$$r = \cos \theta$$

where r represents the sensitivity at angle θ.

Here are some of the cosines of angles that can be found from a calculator:

Angle (degrees)	Cosine
0	1.000
20	0.940
40	0.766
60	0.500
75	0.259
90	0.000
120	−0.500
135	−0.707
160	−0.940
180	−1.000

Now, if these cosine values are plotted out on *polar* graph paper, as in the simplified version below, it can be seen that we can get a figure of eight pattern.

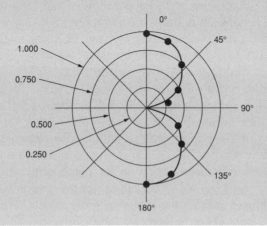

Two important practical points about this type of polar response may be mentioned here:

1. All microphones of this type show a phenomenon known as the *proximity effect*, sometimes called *bass tip-up*. This means that their bass output is increased when the source of sound waves is close to the microphone. The amount of this effect may vary with different microphones but speech typically begins to sound 'bassy' when the speaker is closer than 30–50 cm.

   Bass tip-up is actually used to good effect in one type of commentator's microphone called a 'lip ribbon' microphone. Held close to the mouth, the bass tip-up would be excessive if it weren't compensated for by a bass-cut system. In cutting down the unwanted bass from a close sound, distant bass sounds (crowd noises at a football match, for instance) are greatly reduced.
2. Another and undesirable effect is that these microphones are *rumble-prone* – any movement of the microphone is likely to result in excessive rumbling noises in the output.

These microphones are also extremely sensitive to air movements such as wind.

3. Cardioids

Mathematically a cardioid pattern may be represented by

$$r = 1 + \cos \theta.$$

Notice that this is the equation for a figure-of-eight with the addition of a constant, namely 1. In acoustic terms this means combining a figure-of-eight microphone with an omnidirectional microphone (for which the output is constant in the sense that it does not vary with angle). Some early cardioid microphones actually used two separate units, one omni and one figure-of-eight, in the same casing, but they were large and heavy and their cardioid patterns were usually poor.

Modern cardioids use a different technique, known as the *phase-shift principle*, in which some sound is allowed to reach the back of the diaphragm by means of small slots or holes in the microphone casing. An acoustic labyrinth is used to delay slightly sounds entering via these apertures, and the overall effect is that 0° sounds have what is in effect a fairly large pressure gradient action on the diaphragm, whereas those arriving from the rear (180°) and travelling via the apertures, strike the back of the diaphragm at the same instant as those which have diffracted round to hit the front. There is then no overall force on the diaphragm.

(In an electrostatic cardioid the backplate is perforated to allow sound waves to get to the rear of the diaphragm.)

Because there is an element of pressure gradient operation in the way they work, cardioid microphones generally show a degree of bass tip-up.

They are frequently described in manufacturers' literature as being 'pressure gradient', which, of course, they are, at least in part.

4. Hypercardioids

The construction of these may be thought of as being based on the idea of a cardioid but with freer entry of sounds through the apertures, making the microphone's behaviour nearer to that of a figure-of-eight. This combination of part cardioid, part figure-of-eight can be made to be hypercardioid. Bass tip-up is, not surprisingly, usually more pronounced with hypercardioid microphones than cardioids.

5. 'Gun' microphones

These use what is called an *interference tube* in front of the diaphragm. As mentioned earlier there are two widely encountered lengths. In each case the tube is slotted or perforated along the side. The principle is that sounds which arrive off the axis of the tube enter at many different points along its length and therefore reach the diaphragm at different times. This results in some cancellation as the peaks of some waves will coincide with the troughs of others – provided the wavelengths are short enough for this to happen. This is why such microphones are only slightly directional at low frequencies. In fact, they would be omnidirectional but it is common practice to make the basic capsule a cardioid so that some directional effects remain even at these bass frequencies.

6. Boundary microphones

If a microphone is positioned so that its diaphragm is a short distance above a hard, reflective surface, there are likely to be undesirable effects caused by the diaphragm receiving two sets of sound waves: direct ones and those reflected from the surface. The reason for this is that if there

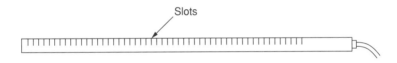

Figure 4.10 Sketch of a gun microphone

happens to be half a wavelength difference between any components of these two sets of waves then there will be at least partial cancellation of these particular frequencies. Equally, there may be unwanted reinforcement of other frequencies. The result is an inaccurate sound pick-up.

In boundary microphones, however, the microphone is mounted so that the diaphragm is to all intents and purposes in the plane of the reflective surface, or is placed so closely above it that cancellation effects, if they occur, will be above the audible frequency range. In either case, the reflective surface adjacent to the diaphragm must, in a practical and portable microphone, be fairly small – 20 cm diameter is typical – but when placed on the floor or on a table the effective surface now becomes much larger. It is possible to construct a somewhat makeshift but a perfectly usable boundary microphone by simply taping a small conventional microphone, preferably of the personal type because of their small size, to a large, flat, reflective surface, such as a table top. The microphone body should be parallel with the table top so that the diaphragm part is as close to the table as possible but not obscured in any way.

Sensitivities of microphones

It is often important to know the *sensitivity* of a particular microphone – that is, the electrical output for a given sound pressure incident upon it. Of course, any microphone which is in use will be subject to a constantly fluctuating set of pressures, but it is still possible to set out approximate indications. Manufacturers generally adopt any one of about three sets of figures:

1. The output in dB relative to, usually, 1 volt for a sound pressure of 1 pascal (1 V/Pa) (column 1 in Table 4.2).
2. The output in millivolts for a sound pressure of 1 μbar (column 2).
3. The output in millivolts for a sound pressure of 1 Pa (1 Pa is the same as 10 μbar and many manufacturers use 10 μbar rather than 1 Pa) (column 3).

All of this is confusing. Fortunately, though, it is often *relative* sensitivities which are important: is this microphone more or less sensitive than that one and by roughly how much? Table 4.2 compares the three sets of sensitivities most commonly used.

Conversion from voltages to decibels can be done by the formula given in Chapter 1, Part 2. For example, if a microphone is described as producing an output of 1 mV (one thousandth of a volt), the conversion is:

$$20 \log (1/1000) = 20 \times (-3) = -60 \text{ dB}$$

Table 4.2

dB relative to 1 V/Pa	mV/μbar	mV/10 μbar = mV/Pa	Approximate rating
−20	9.5	95	
−25	5.5	55	
−30	3.0	30	very sensitive
−35	1.8	18	
−40	1.0	10	fairly sensitive
−45	0.55	5.5	
−50	0.3	3.0	medium
−55	0.18	1.8	
−60	0.10	1.0	insensitive

Phantom power

An electrostatic microphone needs, as we have said, a source of power to operate the preamplifier and provide a voltage across the capsule. (Electret microphones don't need a voltage in the capsule but they still need power for the preamplifier.) The power consumption with modern microphones is very small. For example, a well-known type of 'tie-clip' micro-phone, frequently worn by television presenters, can be operated from a single 1.5 V battery, the life of which is likely to be up to 5000 hours – equivalent to 8 hours a day for more than 18 months!

On the other hand, it may not be convenient always to use batteries and very many sound mixing units – almost all professional ones – and also good portable audio recorders contain circuitry which feeds the necessary power down the microphone cable. The way in which this is done is shown in Figure 4.11.

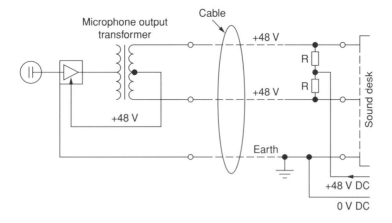

Figure 4.11 Phantom power

The power unit, which is remote from the microphone, provides 48 V d.c. (but note that there are other systems; see below), this being an accepted worldwide standard. The positive (+) output is 'shared' between the two signal-carrying wires of the microphone cable, while the important earth wire, which is usually in the form of braiding to give electrical screening of the signal wires, is connected to the negative (−) part of the supply. (This negative part can be described as being at 0 V.)

The vital point to realize is that *both* signal wires are at the same electrical potential and there is therefore no voltage between them. Consequently, almost any type of microphone could be connected to the system without harm − but there is an important exception which will be dealt with under the heading of 'Balanced wiring' (see below). The resistors R are of fairly high value − 6.8 kohm (kΩ) being preferred. This is high enough to limit the current in the event of an accidental short circuit.

Other powering systems exist, as we have said, but their use tends to be specialized and limited to a small range of microphones. The one described above is known variously as '48 volt phantom' or 'standard phantom' powering.

The term 'phantom' appears to be derived from a system of telephony when three circuits were carried on two pairs of wires − one pair carrying one 'leg' of the third circuit and the other pair carrying the remaining 'leg' in a manner not unlike the powering system we have described. Because one of the telephone circuits had no separate physical existence of its own it was known as a 'phantom' circuit.

Balanced wiring

It is vitally important in things like microphone cables, where the voltages are very small, that all possible precautions are taken to minimize the risks of interference caused by mains hum and other sources being induced into the cable. Careful screening, mentioned above, obviously helps, but a major method of protection is to have both signal wires as similar to each other as possible. In brief, this means that any induced interference voltage is the same in each wire and can thus be 'cancelled out'. Such an arrangement is called *balanced wiring*.

A particularly good example of this sort of cable is what is known as *star-quad*. There are four cables inside a screened outer. Imagine that the viewer is looking at the open end: they would see the four cables arranged in a sort of square. Opposite wires are connected together, and the four are twisted to make a spiral. This makes the arrangement as electrically balanced as possible.

Most domestic audio equipment is *unbalanced* and this can be quite satisfactory for short cable runs – up to a metre perhaps – but unbalanced microphone cables any longer than this are likely to pick up unacceptable amounts of hum and possibly other interference. This topic is dealt with more fully in Chapter 8, Part 2.

Linking together professional balanced equipment with non-professional unbalanced units is almost certainly going to be fraught with problems in the form of unacceptable hum – and maybe other unwanted effects like distortion. The balanced/unbalanced situation can often be cured with a suitable transformer, but this MUST be of the kind that gives complete isolation. What are termed 'through earths' – where an earth connection links primary and secondary – must be avoided.

It can safely be said that all professional microphones are provided with balanced outputs. Non-professional ones may not be.

There are obvious dangers in connecting non-balanced microphones to equipment providing phantom power. Because of the high-value resistors there is unlikely to be any damage, but it will probably mean that the microphone, especially if it is an electrostatic one, will not work.

It may be of interest to mention that television studios are regions of great hazard as sources of interference. In particular, the dimmer circuits of the lighting systems radiate large amounts of the harmonics of mains frequency. This problem has, in recent years, extended to even quite small halls and theatres, where 'affordable' electronic dimmer systems have become quite common. In a domestic environment fluorescent lights and wall-mounted dimmer units can be a frequent cause of trouble.

Radio microphone data

There is considerable variation between models and the following is no more than a rough guide:

Frequency	around 174 MHz
Transmitter power	2–20 mW
Transmitter battery life	4–5 hours
Range	10–30 m, depending on circumstances. (more than 30 m is possible line-of-sight out of doors)

Almost all make use of an f.m. system (= frequency modulation. The *carrier frequency* is made to vary in proportion to the microphone output voltage.)

A danger with radio microphones is that a performer with the microphone/transmitter who moves about a lot may enter a region where a reflection of the radio signal off a sufficiently large metal surface may cancel the direct signal at the receiver. When possible, all the positions in which the person may be should be checked in advance, although this is not always practicable.

The risk of radio black-spots can be greatly reduced by the use of a *diversity receiver system*. The latter has two receiving aerials spaced a short distance apart – perhaps a couple of metres – and the electronics in the unit continually monitor the outputs of the two aerials, switching instantly and silently to the one with the stronger signal. Diversity units are expensive but can be hired.

It may do no harm here to repeat the warning near the end of Part 1 of this chapter: in the UK, several frequencies are allocated for general use, although the equipment must be approved by the DTI (Department of Trade and Industry). This means that there may be other people in the vicinity operating radio microphones quite legitimately on the same frequency. Make enquiries beforehand if possible and be prepared to negotiate!

Questions

1. Phantom power is delivered to a microphone that needs it via
   a. Special three-core cable   b. Standard three-core microphone cable
   c. Three-core mains cable   d. Two-core cable

2. Which types of microphone are likely to need phantom power?
   a. All types   b. Ribbon
   c. Moving coil   d. Electrostatic

5 Using microphones

Objectives and problems in recording

It is obvious that any reproduced recording should aim to be either a replica of the original sound or, by means of mixing and editing techniques, a combination of original sounds (not necessarily natural ones) that achieves a particular artistic result. These, then, could be described as the general objectives. The problem is really how to achieve them. The correct use of microphones will go a long way towards realizing this aim, but what in practice is meant by 'correct'?

Sensible positioning of the microphones is very important, for a start. Experience helps to bring good results quickly and trial-and-error may also be satisfactory – eventually! And while there are no hard and fast rules and no real short cuts, perhaps this chapter can help to point the reader in the right direction. A good way to start may be to highlight some of the more serious errors that can occur.

1. Overload distortion

Typically caused by excessive sound levels at the microphone, this is particularly objectionable as the sound waveform is 'squared off'. It's important to realize, though, that it is rare for distortion of this type to occur at the microphone diaphragm. Almost invariably it happens at the first amplifier in the chain, which in the case of electrostatic microphones is the pre-amp.

Overload distortion is best avoided by checking on the quality of the recording, either by a playback of a test recording, and/or by keeping a careful watch on monitoring instruments in the recording equipment. (But be careful: not all monitoring meters are reliable in this respect. We explain this point in Chapter 6 when dealing with visual monitoring.)

There should be little risk of this kind of distortion if the microphone is no closer than about 20 cm from a person speaking. Musical instruments may need greater working distances, especially things like trumpets and almost all other brass instruments.

2. Breath and other wind effects on the diaphragm

These show up as a disturbing 'blast' effect on playback, sometimes more noticeable on 'p' and 'b' and similar 'explosive' sounds in speech. Microphones differ widely in their susceptibility to this effect – some can be used almost touching the mouth of a singer, others can be affected by a gentle breeze out of doors. One answer is to have a suitable *wind-shield* which may be supplied with the microphone as an accessory or may be integral with it. Discs of gauze mounted a few centimetres in front of the mouth are often used as windshields for vocalists.

An often acceptable ad hoc remedy is to wrap a few thicknesses of cloth, such as a clean handkerchief, round the diaphragm end of the microphone and hold this in place with a rubber band. It is worth noting that the apparent density of a makeshift windshield is not always an indication of its effectiveness. In general, the larger the windshield the better – a light spherical gauze shield of large diameter may be far more use than a closely wrapped cloth and it will probably have a less detrimental effect on the sound quality.

Whatever the windshield though, windy conditions out of doors should be avoided if at all possible. If recordings *have* to be made in circumstances such as these, then working in the downwind side of a building may solve or at least reduce the problem.

3. Rumble

This, if transmitted through floors or tables and travelling thence to the microphone stand, can be very distracting. There may be quite innocent causes – the gentle tapping of a possibly nervous foot against a table leg for instance, and it is quite possible for the offender to be unaware of this at the time. The effect will usually be shown up on loudspeaker or headphone monitoring and if possible the noise should be stopped at source. Failing that, a suitable support for the microphone stand can help, but it may be little use just putting a rubber sheet underneath the stand. Any pad should be thick enough and compliant enough for the microphone and stand to sink into it slightly, as explained in Chapter 3 under 'Structure-borne sound'.

4. Excessive background noise

It is a curiosity of the hearing process that one is often unaware of extraneous noises until a recording is played back, when it is probably too late to do anything much about them. This is at least partly an aspect of the cocktail party effect mentioned in Chapter 2.

Obviously the best remedies are either to stop the noise or find another and quieter location. If neither is possible, then the use of a directional microphone, such as a cardioid with its 'dead' side pointing towards the noise source, can be tried, together with having the microphone as close as possible to the wanted sound, consistent with avoiding distortion and breath noises.

Note, however, that some background noise may be important in establishing the environment. For instance, a sound commentary for a videotape about an industrial process probably should have some of the relevant noise in the background, certainly if the presenter is apparently in that environment.

5. Too much reverberation

This is a very common failing. It arises from either having too long a reverberation time in the room and/or having the microphone too far away from the sound source – most often the latter. It is also typical of much camcorder work using the built-in microphone, especially when high magnification zooming is used: the picture suggests closeness to the action but the microphone is some distance away.

This is an appropriate place to go into the topic of *sound perspective* – the apparent distance of the sound source – in this case when it is listened to through loudspeakers. It depends on the ratio of the *direct sound* to the *indirect sound*. By 'direct' we mean the sound which travels directly to the microphone; the 'indirect' sound is that which travels to the microphone having undergone reflections from walls, floor, ceiling and other surfaces – in short, the reverberant sound.

> *Sound perspective* – The apparent distance of a source of sound. It depends on the ratio of the sound which travels directly to the listener (the 'direct sound') compared with the reverberant, or 'indirect' sound. The greater the proportion of indirect sound the further away the source seems.

Briefly, it can be said that too much indirect sound gives speech, to take an example, an excessively distant effect. Usually the aim with recorded speech is to give the impression that the speaker is in the same room as the listeners. If this is not so there can be a sense almost of alienation and lack of any sort of contact with the speaker. This is particularly the case with sound-only recordings. With the medium of television, especially if the speaker is in vision, the brain uses visual information such as facial expressions, lip movements and other gestures to supplement the information reaching the ears. All this means that there can be a little latitude in sound perspectives when there are accompanying pictures. We emphasize the 'little' as large discrepancies between sound perspectives and visual perspectives can be severely distracting.

With any sound source the distance and type of microphone both have a big effect on the perspective. Taking distance first, it must be said that, if in an attempt to reduce the indirect sound, the microphone is brought close to a speaker, there can be attendant problems. For example, if a microphone is 30 cm away from a person's mouth then small (e.g. 5 cm) to-and-fro movements of the head will probably not affect too seriously the level of the sound signal. However, if the working distance is reduced to 10 cm, the same to-and-fro movements could cause most undesirable fluctuations in the signal.

Directional microphones can help to alter the apparent perspective. Compared with an omnidirectional microphone, the rejection of random indirect sound by a cardioid or figure-of-eight microphone is significant. A hypercardioid microphone can be even better.

With excessive reverberation then, it may be a matter of attempting some temporary reduction in reverberation time (see Chapter 3), having the microphone as close as possible to the speaker(s) and using one that is directional. A degree of equalization (see Chapter 8) can often reduce the worst effects of reverberation. Again, trial recordings and a carefully listened-to playback are essential.

It may be in order here to make a comment about playbacks. There is a very common, and perhaps natural, tendency for amateur performers, on hearing their spoken or musical efforts, to react with a mixture of embarrassment and delight. This is manifested by various symptoms, ranging from stunned silence to shrieks of laughter. Whatever the reaction, the sound recordist who is seeking to produce as good a result as possible must never allow these effects to interfere with his or her work. If necessary, playbacks used to assess the quality of the recording must be heard in the absence of any performers.

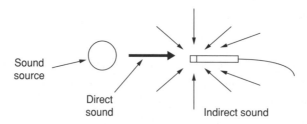

Figure 5.1 Direct and indirect sound

It is nevertheless often a good thing to let, for instance, amateur actors hear themselves at some point to help them eradicate mannerisms, maybe boost their confidence, or perhaps in some cases to make them realize that they aren't quite as good as they thought. A little practical psychology is often necessary!

6. Incorrect sound perspectives with more than one sound source

As an example, a discussion between two people should normally sound to the listener as if both are the same distance away. This means, if a single microphone is being used, that they are equal distances from it. This can be difficult if one has a much louder voice than the other. In practice, it may be possible to move the louder speaker a little further away from the microphone before he/she begins to sound too distant. Or, if a cardioid microphone is being used, the louder voice may be moved round to the side of the microphone. A better solution is to have more than one microphone, perhaps one for each speaker, so that each can be approximately the same distance from their microphone. The levels are then adjusted on the mixing unit.

The point should be made here that 'correct perspectives' in a recording does not necessarily mean that there should be the same perspective for all sources. The correct perspective in any situation is that which conveys to the listeners the illusion that the producers of the programme wish to achieve. To take an example, a small musical group such as a string quartet, or four 'barber shop' singers, would be expected to be set out on the platform so that each performer is roughly the same distance from the audience. In that case the sound perspectives should be about the same for each performer.

A large orchestra would have the strings closer to the audience than, say, the brass, woodwind or percussion. It would then be in order for the further instruments to sound more distant.

Note that rock music is not being mentioned, but only because there are fewer established conventions than is the case with classical music.

The basic question is, what would the listeners expect? If the recorded perspectives conform to that expectation, at least within reasonable limits, then all is well. If there are no particular expectations, then perhaps either perspectives do not matter or they can be whatever the producer/ musicians want.

In drama, different sound perspectives can be vitally important in establishing the scene – a person who is supposed to have entered a room from a door in the far wall should sound much further away than people in the foreground, but intelligibility must not suffer in the process. Compromise is inevitable.

7. The balance is wrong

This means that the relative loudnesses of the various sources, as heard on playback, are inappropriate. It is quite impossible to lay down rules – it all depends on what effect the producer/recordist wishes to obtain. To give a few examples: the speakers in a round-table discussion should all sound more or less equally loud; a singer with a piano accompaniment should normally be a little louder than the piano, although much can depend on the music; the sections of a choir should usually be of comparable loudness, at least as far as the setting-up of the equipment goes – it will be the job of the conductor to vary the balance within the choir according to the music being sung.

When several performers are being recorded, conflicts between balance and perspectives are more difficult to resolve if only one or two microphones are being used. It may then be better to have a number of microphones and achieve the desired balance at the mixing stage.

Specific applications of microphones

As we have said, there are no hard and fast rules about the use of microphones. All too often a recordist has little or no choice in what microphones are available to him or her. Skilled recording frequently depends on the intelligent use of what comes to hand. Nevertheless, the following can be regarded as guidelines. While the list is short it covers sufficient examples for the principles to be easily adapted to other situations. The notes above, about errors to avoid, should be borne in mind.

At the moment it is assumed that the recording is in mono. What is set out below may need mental revision by the reader when the chapter on stereo has been read. In general, though, any modifications to these guidelines will be small when stereo is being used.

1. Interviews out of doors

A single omnidirectional microphone with a windshield and held by the interviewer ought to be sufficient. A good interviewer will hold the microphone so that it is roughly equidistant from both mouths but if necessary holding it closer to the weaker voice.

There should be no need to move the microphone to-and-fro depending on who is speaking unless they are widely separated or there is considerable background noise and it is necessary to pick up as little of the noise as possible. (It's often valuable to watch the techniques of television interviewers.)

Moving coil microphones are usually to be recommended – electrostatic microphones, as we have said, tend to suffer from humidity problems unless they are of the r.f. type mentioned in Chapter 4.

2. The solo presenter

We will take the case of the presenter who may be in front of a camera. A good solution is the small 'tie-clip' microphone. The best of these are costly, but relatively inexpensive ones may be perfectly adequate. An alternative is for the presenter to have a hand-held microphone. The latter may be inconvenient if he or she has to hold a clipboard for a script or notes. On the other hand, in front of a camera, a hand-held microphone might sometimes be the answer for a presenter who, like some amateur actors, does not know what to do with his or her hands!

3. Piano

This is a notoriously difficult instrument to record really well. Assuming pessimistically that the ultimate in quality is not going to be achieved anyway, the following notes may be helpful.

Microphones ought not to be too close to the piano. It is important that the microphone can 'see' the whole instrument, as otherwise the balance between different sections of the strings may be wrong. For a classical piano piece, the microphone can be relatively distant, so that there is some reverberation 'surrounding' the instrument in the recording. For a grand piano with the lid raised, one usually good position is 2 or 3 metres from the piano and in line with the treble end of the keyboard, so that the lid acts as a reflector towards the microphone. This assumes fairly good acoustics, though. In more popular music, microphones may have to be close to the piano to avoid picking up other instruments and then it may be necessary to have two or more microphones to cover each section of the strings.

Upright pianos vary considerably. Sometimes it can help to raise the lid, or remove a panel. Also, a lot of sound is radiated from the rear so that a good place for the microphone(s) may be behind the instrument (but not if the piano is close to a wall).

4. A fairly reliable general rule

While we have said that there are no hard and fast rules about microphone placing, there is nevertheless one very good general rule to adopt when 'miking' musical instruments – if an accurate recording of the true sound of that instrument is wanted.

It is to place the microphones so that they are looking in the direction that an audience would. In other words, the microphone takes the place of the audience, but for reasons mentioned earlier will almost invariably have to be much closer to the instrument.

Following this general rule, and to take one or two examples as illustrations, a clarinet is usually pointed downwards by the player. A microphone should not be directed into the bell of the clarinet but should be in front, 'looking' at the entire instrument.

On the other hand, trumpets and trombones are usually aimed at the audience and therefore microphones can look towards the bell (but not too close if a harsh 'breathing' quality is to be avoided).

The bell of a French horn points *away* from the audience, frequently towards a reflecting wall or other hard surface, and any microphone should preferably be looking towards the wall as seen by the audience. And so on.

5. Small singing group

Assuming reasonable acoustics, this is one case where an omnidirectional microphone can be used. The other assumption is that the group is *internally balanced* – that each member's voice is at the right strength compared with the others. If this is not the case then there are three options:

1. Use more than one microphone and adjust the balance on the mixer.
2. Have the quieter voices nearer the microphone.
3. The conductor or leader should try to get the loud singer(s) to sing more quietly. The latter is probably the best solution, but in practice may be the most difficult!

Much depends, of course, on whether the group is performing to an audience or just to the microphones. Almost always, musicians work better with an actual audience – there can be some kind of feedback which stimulates the performers into 'giving of their best'. The same is true of actors. Only well-trained and experienced actors play really well to inanimate things like microphones!

6. Discussions

Much of what has been said above about small singing groups applies to discussion groups. However, it is likely that clarity and intelligibility are of prime importance in a discussion – the niceties of balance and perspective will probably be secondary. An omnidirectional microphone in the centre may be satisfactory if no speaker is too far from it, and/or the acoustics are relatively 'dead' (the RT is short).

A most important point to make here is that no microphone should ever be placed a short distance (5–10 cm, say) above a hard table top. The reason is that some sound waves travel directly to the diaphragm and others arrive after being reflected from the table. The latter will have travelled a greater distance and this can mean that some wavelengths will reinforce, others will cancel, resulting in a peculiar sound quality. Boundary microphones (see Chapter 4) avoid this problem.

7. Small musical group with singer (classical)

If the group is accustomed to public performances it may be necessary to do no more than place the microphone in front of them, the distance being determined by the size of the group, the polar diagram of the microphone and the acoustics of the hall. The exact position should be determined by listening critically to test recordings or by real-time monitoring.

The same technique should work adequately with a larger choir or orchestra.

8. Small musical group with singer (rock)

The main problem here is that most of the instruments will have amplifiers and be very loud, whereas rock singers are not noted for the loudness of their voices. The singer must then have a microphone of the type which can be held close to the mouth and amplified independently of the microphones in the band. Individual microphones for the instruments are likely to be needed, maybe in front of the instrument's loudspeaker. An alternative is to use *direct injection (DI) boxes*. These allow an electrical signal from each instrument to be routed directly to the mixer.

DI boxes can generally be hired if they are not otherwise available, but they should be selected with caution as poor electrical insulation in the instrument's amplifier can result in dangerous voltages reaching the mixer, or worse still, a performer.

Finally, in this chapter, a few important cautionary words.
First **SAFETY**.
Cables must always be laid out on floors where there is no chance of anyone tripping over them. At doorways they should be carried over the top if at all possible, or else they can be laid underneath mats or carpets. If they *have* to go on a bare floor with no covering, they should

be securely taped to the floor with no loops that anyone can catch with a foot.

All other items of equipment – loudspeakers for instance – must be securely rigged so that there is no chance of their falling and hitting someone. Cables from microphones on a table or fixed to a microphone stand should be taped to the bottom of the stand, or the bottom of a table leg, so that the cable then lies as close to the floor as possible, again taped down if appropriate. All of this is important even if there is only a small group of performers. It is even more important when there is an audience.

It is a very good idea to check that there is adequate public liability insurance. It may not be your responsibility to deal with insurances but be prepared to ask the question.

Safety is such an important topic that much of what is set out here is repeated in Chapter 14.

Lastly, and this is nothing to do directly with safety, the sound operator must remember that he/she must do nothing to spoil an audience's enjoyment. To begin with, microphones and their cables should be as inconspicuous and tidy as possible. The public have come to see a performance and they don't want an exhibition of the operator's equipment. The ideal sound operator is transparent!

Questions

1. Which of the following microphone types may show the proximity effect ('bass tip-up') when a performer is close to the microphone?
    a. Omnidirectional   b. Cardioid
    c. Figure-of-eight   d. Hypercardioid

2. A generally good rule about placing microphones when high-quality pick-up of a musical instrument is required is:
    a. Place the microphone above the instrument so that it is 'looking down' at it
    b. Place the microphone so that it is 'looking' in the same direction that an audience would do
    c. Have the microphone as close as possible to the instrument
    d. With brass instruments have the microphone 'looking' into the bell of the instrument

6 Monitoring
Part 1

Monitoring in an audio context means keeping a careful check on the sound signal. There are two broad requirements. One, *visual monitoring*, can be regarded as technical – that the signal voltage keeps within prescribed limits for reasons which will be explained. The second, *aural monitoring*, uses the operator's ears not only to listen for technical imperfections which cannot be shown easily (if at all) by visual means, but also to fulfil what is at least partly an artistic function – that the balance, perspective, performance and so on are satisfactory. We will take these in turn.

Technical monitoring

To understand properly the need for this, we must first consider *dynamic ranges* – the difference in dB between the loudest and the quietest signals. Of all the likely sound sources, a symphony orchestra probably has one of the widest dynamic ranges. Table 2.1 in Chapter 2 shows that an orchestra produces a maximum sound level in excess of 100–120 dB(A) at peaks – but there may well be bars in the music which require all instruments to stop playing and the sound level in a quiet studio may then be less than 20 dB(A). The dynamic range of an orchestra is thus likely to be about 100 dB – possibly even more.

DEFINITION
Dynamic range is the difference in dB between the loudest and the quietest audio signals.

The dynamic ranges of other programme sources are generally much less. Normal speech, for instance, may well be contained within 30 dB.

It is important to remember that all items of audio equipment – microphones, recording machines, amplifiers, loudspeakers and so on – have their own electrical limitations. If the signal level is too high, there will be distortion because the equipment cannot handle these voltages – and there may even be a risk of damage in some cases. At the other end of the scale, there is a background hiss which arises because, to take one reason, electricity consists of individual particles – electrons – and is not a smooth stream of something like water. While we can generally pretend that electricity is a continuous fluid, because light bulbs do not flicker because of the particle nature of the current, when it comes to very small currents which are going to be amplified the fact that electrons are not continuous can begin to show up. This is a matter of great importance, for example, in detecting the faint signals in radio astronomy, and special techniques have to be adopted.

With analogue magnetic recording there is another source of hiss due to the fact that the magnetic material on the tape is made up of particles of magnetic material – particles again! With digital recording systems, as we shall see, the effects of tape hiss are eliminated.

The dynamic range, or *signal-to-noise ratio*, which we shall take to be the same thing, on conventional audio cassettes depends on the tape quality, but 50 dB is a reasonable figure, if perhaps a little optimistic. As we explain later, this can be extended by *noise reduction systems*, but the fact remains that the range is very much less than that needed for a symphony orchestra. Even a compact disc's dynamic range is not infinite, but being in the region of 90–100 dB is probably good enough for all practical purposes.

If we think about rock music, although the dynamic range may be fairly small (from very loud at minimum to exceedingly loud at maximum!), there is still the problem of avoiding high signal levels which could cause distortion.

How the dynamic range of a sound source is reduced to 'fit' in the signal-to-noise ratio of the system is something that we shall deal with later, but it is enough at the moment to realize that it is vitally important for any operator to know whether or not the signal being dealt with is within the limits of the system.

Of course, there are two limits, an upper and a lower, and of the two the upper one is the more important for the simple reason that this represents the loud end and if distortion occurs it is going to be very apparent and unpleasant. Therefore, a means of detecting the programme peaks is crucial.

Let us say straight away that the ear is useless for this purpose. To begin with, the sound level coming out of a loudspeaker is affected by the setting of the volume control. Secondly, the ear adjusts to sound levels and what might seem loud to start with becomes a more comfortable level after a few minutes. Furthermore, our judgements of loudness tend to be affected by whether or not we like the music or are interested in the content of the programme. Folk music at a very moderate volume can be excruciatingly loud to someone whose taste is for classical music.

There is thus a need for some kind of electrical measuring device.

There are two broad categories of measuring device that are in widespread use. They are outlined here, but more technical information is given in Part 2 of the chapter.

1. More or less conventional voltmeters

These may not look like voltmeters because of their possibly unusual scales. The best-known example is the *VU meter*, VU standing for volume unit. This is found on much professional equipment and a form of it, with perhaps less rigid specifications, may be seen on the slightly more expensive domestic cassette recorders, although the latter are now more commonly fitted with LED indicating systems (see section 3 below).

The VU meter is reliable and relatively cheap, but it has a major drawback in that it indicates *average* voltages, whereas it is the *peaks* that are the really important things to be monitored, as we have shown.

2. Peak indicating meters

Basically, these are devices with electronic circuitry which detects and holds the voltage peaks long enough for them to be registered by the meter pointer, or some other indicator. The best-known example, at least in the professional field, is the *Peak Programme Meter (PPM)*.

Because the circuitry is relatively expensive (particularly so in the case of stereo versions, which have two pointers with concentric spindles), these are generally only found in fairly costly equipment. However, if equipment is to be hired or borrowed, that which is equipped with PPMs should be preferred.

Figure 6.1 VU meter (courtesy SIFAM Ltd)

Figure 6.2 PPM (Peak Programme Meter) (courtesy SIFAM Ltd)

The true PPM has a black scale with white numerals and the pointer is controlled in such a way that it has a very rapid rise time and a much slower fall-back, these characteristics making it easier to read the signal peaks but adding, of course, to the cost. Considered all round, the PPM is much superior to the VU meter, as it can indicate with adequate accuracy the all-important signal peaks and thus warn of impending distortion.

VU Meter – Essentially a meter which shows average programme voltages.
PPM – A meter which indicates the peak signal voltages. The scale is black with a white pointer, which has a rapid rise and slow decay.

---

3. LED indicators

Coming really into category 2 above, but dealt with separately here for convenience, is a range of relatively low-cost devices using LEDs (light-emitting diodes). These are much less accurate than PPMs, or even VU meters, but they can be made to give a fair indication of signal peaks using what is sometimes called the 'bouncing ball' technique. The electronics associated with the display allow the highest LED segment to stay lit for a second or so after the peak has passed. This can be a very adequate means of monitoring audio signals, although it is much less precise than a PPM. PPMs can be read to an accuracy of 0.5 dB with a little practice, whereas the individual segments of an LED unit are frequently arranged in steps of 2 or even 3 dB. For many purposes, though, this may not matter. I have to say that, personally, I prefer to monitor a recording with a peak indicating LED display rather than with a VU meter, assuming no PPMs were available.

Aural monitoring

For a human being's ears to assess the quality of a sound signal, there needs to be, first of all, good quality listening equipment – headphones or loudspeakers. To begin with, let us state that of the two, loudspeakers are considered by almost all experienced sound engineers to give much better indications of quality than headphones.

The reasons for this are not always very clear. One thing is certainly true: stereo image positions given by headphones do not in general correspond at all well with those produced by loudspeakers – and that loudspeaker stereo images can relate well to the sound sources in front of the microphones.

Of course, it has to be admitted that there are many situations where loudspeakers cannot be used and therefore headphones are the only practical alternative – outdoor location work being one example. One other drawback with headphones is that they tend to isolate the wearer, not just acoustically (which might not always be a bad thing) but also psychologically, from other members of the team. Apart from all else, in studio-type situations the headphone wearer may find it difficult to hear comments or instructions from others, although 'in-ear monitoring' with small earpieces avoids this problem.

We will begin, then, by considering loudspeakers.

Loudspeakers

The main requirements of a loudspeaker are that it can handle with equal impartiality all the frequencies in the region from 30 Hz to about 16 kHz. It should also be able to produce relatively high sound levels if necessary. Notice that the low frequency end is given here as 30 Hz not the 16 Hz stated in Chapter 2 as being the lower frequency that the normal ear can detect. The point is that it has been, and to some extent still is, difficult to make audio equipment operate satisfactorily below about 30 Hz. Luckily there are relatively few sounds of any importance in the range 16–30 Hz, large pipe organs being an occasional exception. Furthermore, the human ear/brain combination has the ability to compensate to some extent for missing bass frequencies ('False bass', referred to on p. 23). Consequently, broadcasters and recording companies have, for a long time, accepted that 30 Hz is a reasonable lower limit in practice, although the increasing use of digital recording equipment removes one (but only one) of the obstacles to really deep bass reproduction.

The requirements in terms of sound levels vary with the situation. A professional studio is likely to want loudspeakers that can produce sound levels in the region of 120 dB(A) at 1 m. Such loudspeakers are large and/or expensive. They can also be a severe hearing hazard! For semi-professional and serious amateur work, the requirement can be much more modest. Maximum sound levels of perhaps no more than 80 dB(A) at 1 m are likely to be perfectly adequate, and such levels are well within the range of relatively small and modestly priced units.

Basic loudspeaker construction

A loudspeaker is, in a way, a microphone 'working backwards' and it may not be surprising therefore to find that the most commonly used loudspeakers have features in common with some microphones. In fact, many loudspeaker units can function as microphones and indeed do so in some intercom systems. In place of the microphone's diaphragm, a loudspeaker has a cone which is made to vibrate and thus generate sound waves.

A favoured material in low cost cones is a form of compressed paper. This can be fairly satisfactory but it may not be easy to achieve a good match between two units – essential for stereo reproduction – because of differences in the orientation of the fibres in the material. Modern high-quality loudspeakers have cones made of vacuum moulded plastic – polypropylene being a good example.

In principle, many systems can be used for the loudspeaker transducer. The one most commonly used is a moving coil unit, as shown in Figure 6.3. This is very similar to a moving coil microphone transducer, except that in a loudspeaker the coil and magnet are much larger.

The reason for the popularity of the moving coil unit is that, in conjunction with a suitable cone and appropriate enclosure (see below), it can

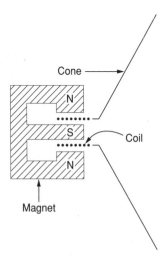

Figure 6.3 Moving coil loudspeaker unit

cover a reasonably wide frequency range and at the same time handle power levels which are adequate for many purposes. This is not to say that a single such unit can meet the requirements set out above, but with good design it is possible to cover almost the full frequency range *and* produce quite high sound levels with two or at most three units in one enclosure.

Other transducer devices besides moving coils find application in, for example, certain 'tweeters' (see below).

The combination of transducer and cone is almost invariably an integral unit. A frequently used term for the combination is 'drive unit'.

Drive unit – Term used to mean the combination of transducer and cone.

The remaining major component in a loudspeaker is the *enclosure* or cabinet in which the drive units are mounted. This is much more than just a box and a good loudspeaker depends heavily on a well-designed enclosure.

To summarize the reasons for the need for a good enclosure, a drive unit on its own radiates sound waves from both the front and rear surfaces of the cone. Imagine that the cone moves forward: a sound wave compression is emitted from the front but a rarefaction (low pressure) wave emanates from the rear. If the wavelength is large, these two sets of waves will diffract round the cone and tend to cancel each other out – the low pressure 'swallowing up' the high pressure. At high frequencies, when the wavelength is small, this diffraction does not occur so that the front and rear waves tend to travel outwards in something approaching straight beams. This means that a listener on the front axis of an isolated drive unit will receive sounds which are seriously lacking in bass frequencies because these have largely cancelled themselves out. Figure 6.4 shows the approximate frequency response of a 25 cm cone unmounted in any kind of enclosure.

A solution, although not necessarily the best one, is to put the drive unit in a sealed box, as shown in Figure 6.5. Obviously, this prevents any sound waves from the back of the cone coming round to the front.

The inside of the box needs to be treated with sound-absorbent material to reduce internal reflections of sound waves, and the box should be made of rigid but non-vibrating material. Chipboard can be satisfactory but birch ply has been found to be excellent in some professional loudspeakers.

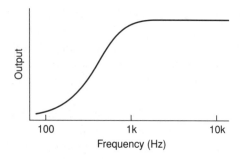

Figure 6.4 Response of an unmounted cone

Finally, it is possible for the complete unit to be compact. A typical volume is around 0.01 m³ (e.g. 30 cm high, 20 cm wide and 15 cm deep).

Unfortunately, sealing the box in this way has an effect similar to holding a finger over the end of a bicycle pump, when it then becomes difficult to operate the pump. In the same way the cone movements, which at low frequencies have to be relatively large to generate the required volume of sound, are restricted. Hence the bass response is then not as good as one might perhaps have hoped for, although it can still be very respectable – and even good enough for some professional purposes.

Figure 6.6 shows a more ambitious design – the *vented enclosure*, sometimes called a 'bass reflex' cabinet. The idea is that the vent and the rest of the cabinet form a device called a *Helmholtz resonator*, designed in this case to have a natural frequency of acoustic vibration in the region of 30 Hz. Careful choice of the dimensions of the unit allow it to be 'tuned' to a frequency which boosts the drive unit's output where it would otherwise be falling off. Vented loudspeakers are usually

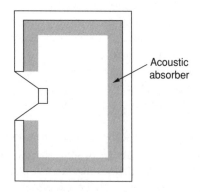

Acoustic
absorber

Figure 6.5 Sealed enclosure loudspeaker

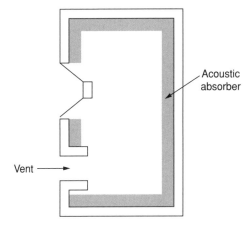

Figure 6.6 Vented loudspeaker enclosure

rather large, heavy and expensive, but they can give both very good sound quality and high sound levels. They are almost invariably the first choice for professional studio monitoring.

Multiple unit loudspeakers

It is impossible for a single drive unit to cover the full audio range and at the same time produce high sound levels. Designers therefore resort to the use of more than one unit in the cabinet. Typically, there is a *bass unit* which covers the frequency range from 30 Hz up to perhaps about 800 Hz; a *mid-frequency unit*, going from there up to around 3 kHz, and finally a *tweeter*, which takes over and extends the range to 15–16 kHz. An electrical circuit called a *crossover network* divides the audio signal into the required bands.

On some high-grade professional monitoring loudspeakers, careful design has made it possible to cover the full range using only two drive units. This simplifies greatly the design (and cost) of the crossover network.

A drawback with large multiple-unit loudspeakers is that, because of the spacing of the units, an operator has to be a little distance – perhaps a couple of metres – away from the loudspeaker. Sometimes this may not be a problem, but when space is limited loudspeakers with minimal spacing of the units are needed so that a good balance between the different sound components is achieved quite near to the loudspeaker. These are often referred to as *near-field* monitors.

Finally, it may be worth pointing out that the sound one hears from a loudspeaker is very much influenced by the listening environment, and it can be argued that this environment should be similar to the final 'customer's' listening conditions. This is obviously an impossible requirement as there are going to be almost as many different such conditions as there are customers. Nevertheless, a step in this direction has been taken by a number of manufacturers of television broadcasting sound mixing consoles. They have incorporated a small *domestic listening* speaker. The programme sound can be switched away from the main high-quality units to these small ones, and this can give an indication of the sound which will be emitted from a typical domestic television set.

Headphones

As far as their construction goes, it is simplest to regard these as miniature loudspeakers. Many, but not all, use moving coil units and in a way headphone design is easier than for loudspeakers because the power requirements are less. In other words, a single unit can produce adequate sound levels and a wide frequency range at the same time – which is not to say that they all do! Choice is very much a matter of cost (headphones being vastly cheaper than loudspeakers of corresponding quality) and comfort.

Listening tests

We will end this chapter with a topic whose importance cannot be stressed too greatly. Any listening test designed to make a comparison between two items of equipment *must be done by direct switching between the two* – what is often termed *A–B comparison.*

The point is that the ear's memory for really accurate recall is short: 1–2 seconds. It is quite impossible therefore to listen to, for example, a good loudspeaker on Monday and then decide that another good one, heard on Tuesday, is better or worse. This kind of comparison may well be feasible if one loudspeaker is good and the other is poor, but assuming that the two are of comparable quality then immediate comparison is the only answer. And this applies to amplifiers, cassette machines, microphones, CD players, the lot. A moment's thought shows that A–B tests are sometimes not very easy to arrange. How would we compare two CD players? We'd have to make sure that each had the same edition of the disc on it (sometimes a record company issues more than one version of the same original; the differences may be very slight indeed, but they might be there). Then the players would have to be connected to the same amplifiers and thence to the same speakers. The discs would need to be cued up so that they are running as closely in step as possible. A switch, preferably clickless, is needed to switch one CD player at a time to the amplifier. And to finish off, the listeners ought not to be aware of which CD player is in use at any one time; the switching must be done randomly and preferably with more than one type of music. Only then can one make a proper judgement! If there are more than two items to be compared, these must be done in pairs. Attempting to make comparison between three or more sounds is very difficult indeed.

The reader is advised to be very wary of statements like 'This amplifier doesn't sound as good as the one I heard at so-and-so's house last week'. Sadly, one has found this sort of thing in the hi-fi press! And I have to say that I'm very sceptical about reviews which talk about the 'musicality' of equipment – amplifiers, CD players and so on.

Questions

1. Which statement is correct about VU meters? They indicate
    a. Loudness
    b. Sound power
    c. Average signal voltages
    d. Peak signal voltages

2. Which statement is correct about PPMs? They indicate
    a. Loudness
    b. Sound power
    c. Average signal voltages
    d. Peak signal voltages

3. A–B comparisons of two separate sound devices are necessary because the ear's memory for accurate recall of sounds is around
    a. 1 ms
    b. 100 ms
    c. 1–2 seconds
    d. 10–20 seconds

6 Monitoring
Part 2

Reference voltages in audio signals

We wrote in Part 1 of this chapter about electrical measurements using VU meters and PPMs, but avoided any mention of the units to be used. This matter should be cleared up before we go any further.

We don't want to make the following part heavy going for those readers who are 'serious amateurs' in the audio world, such as teachers. It is included because there may be readers who are going to use professional, or even semi-professional, equipment which is fitted with decent metering, and to make proper use of it the information here may be very helpful. It might be a good idea to skim over the next part fairly quickly, but be prepared to read it more thoroughly if the need arises.

Technical monitoring could use volts, or millivolts, or any other electrical unit. However, it is generally most convenient to use a decibel scale. That raises the question, 'Decibels relative to what?' A zero based on the dB(A) (Chapter 2) would be impracticable because the dB(A) is an *acoustic* unit, whereas here we are concerned with finding an *electrical* basis for measurement.

It has been the more or less universal practice for many years to take as a reference voltage a value of 0.775 V, and call this 0 dB or *zero level*. The number 0.775, which seems curious, in fact has a historical background. In the early days of broadcasting and telephony, both in the UK and elsewhere, a useful standard for audio systems was deemed to be a power of 1 mW (1 milliwatt). This could be handled by quite modest little amplifiers and cables, but was at the same time a high enough power for this level of signal not to be susceptible to interference from adjacent circuitry. A standard resistance (strictly we should say *impedance*) was also adopted, namely 600 ohms, this being apparently the typical value for a pair of conventional telephone wires of considerable length.

Zero level – A standard reference voltage equal to 0.775 V.

It is not very easy to measure powers such as a milliwatt, but voltage measurements are much simpler. Thus, voltage measurements across, in this case, 600 ohms, can be equivalent to power measurements. The reader may well remember that power, voltage and current are related. Here we make use of the fact that power = (voltage)^2/resistance.

Now 1 mW is 0.001 W and the resistance is 600 Ω (Ω means ohms), so inserting these:

$$0.001 = (voltage)^2/resistance$$

$$\text{i.e. } (voltage)^2 = 0.001 \times (resistance)$$
$$\text{or } V^2 = 0.001 \times 600$$
$$\text{so that } V^2 = 0.6$$
$$\text{and } V = \sqrt{0.6}$$
$$\text{making } V = 0.775 \text{ QED}$$

Hence, provided all significant resistances were 600 Ω, simple voltage measurements enabled an engineer to calculate the power if he needed to. In more recent years it has not been necessary to make all important resistances 600 Ω and the notion of 1 mW as a standard power has gradually been dropped. All that remains is the standard voltage of 0.775 V and this is referred to as 'zero level'.

PPMs and VU meter readings

The photograph of a PPM (Figure 6.2) shows that its scale is numbered 1 to 7. These are arbitrary units but are related to the signal level in the way shown in Table 6.1.

It will be seen that there are 4 dB between the divisions and one of the advantages of the PPM is that since pointer deflection is proportional to the level in dBs, it relates approximately to perceived changes in loudness. This is not to say that the PPM is a satisfactory loudness meter – it is not, for the very good reason that it does not make allowances for the way in which the ear's sensitivity varies with frequency.

A VU meter scale (see Figure 6.1) is marked in dB but suffers from the fact that the dB readings are unevenly spaced, being cramped at the lower end. What is often confusing is that the '0' on a VU scale does not represent zero level. The modern convention is that it indicates +4 dB above zero level. Unfortunately, different manufacturers seem to adopt slightly different VU meter standards, so that the user may be advised

Table 6.1 PPM scale markings

PPM	dB relative to zero level	Voltage
1	−12	0.19
2	−8	0.31
3	4	0.49
4	**0 (zero level)**	**0.775**
5	+4	1.23
6	+8	1.96
7	+12	3.08

to let the pointer exceed '0' (into the region where the scale is coloured red) for a proportion of the time.

Some VU meters have two scales, one in dB and the other in percentages, where the '0' is also labelled 100%.

PPM reading '6' is regarded as the maximum signal level which can be allowed and there is a risk of distortion somewhere in the succeeding chain of equipment if '6' is exceeded.

Two things are worth noting:

1. The peak permissible level (i.e. PPM 6) is 8 dB above zero level.
2. The voltage corresponding to this peak is 2.5 times 0.775 V. Or, putting things slightly differently, zero level is 40% of the peak voltage.

Helmholtz resonators

These are particularly interesting acoustic devices and have many uses. They are mentioned here because of their application to vented enclosure loudspeakers. The basic form is shown in Figure 6.7.

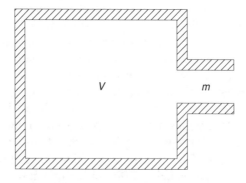

Figure 6.7 Basic Helmholtz resonator

Named after Hermann von Helmholtz (1821–1894), who may have discovered the effect, the action is that the mass (m) of air in the neck corresponds to a weight on a spring, the spring in this case being provided by the *compliance* of the volume of air (V) in the body. Just as a weight on the end of a spring can oscillate up and down, so can the air in the neck oscillate. It might seem that the mass of air in the neck is too small to have any significance, but it must be remembered that the air in the body is a very 'weak' spring.

The resonant frequency of a resonator of this type is given approximately by the formula below. (As always with formulae, consistency in the units is essential. If A is in (metres)^2 then the volume must be in (metres)^3, and so on.)

$$f = 54\sqrt{A/lV}$$

where A is the cross-sectional area of the neck, l is the length of the neck and V is the volume of the body.

Of great value in many aspects of acoustics is the fact that low resonant frequencies can be obtained with Helmholtz resonators of relatively small size. If there is some substance in the neck which can provide enough friction to the air movements (gauze for instance), the device becomes a sound absorber at its resonant frequency. If there is only a little friction then it can re-radiate sound. Clearly, getting the right amount of friction in the neck is critical in designing a good vented loudspeaker.

Gin and whisky bottles (empty of course!) are good Helmholtz resonators. Blowing across the neck excites the natural resonant frequency, which is typically around 110 Hz for 70 cl bottles. The scientifically minded reader who measures the internal dimensions of a whisky bottle and then applies the formula given above mustn't be too surprised if calculation fails to agree with any measurements. One reason is that the acoustic length of a tube – in this case the bottle's neck – is always greater than the physical length!

Loudspeaker power

There is an unfortunate tendency for those who take advertisements at their face value to believe that the higher the power rating of a loudspeaker, the better the quality of reproduction from it. Thus, it might be supposed by some that a 100 W loudspeaker is twice as good as one rated at 50 W. The simple fact is that the only information the power

Table 6.2 Specifications of an acceptably good small loudspeaker

Height	300 mm
Width	200 mm
Depth	160 mm
Frequency response	60 Hz to 15 kHz
Maximum input power	25 W
Max. sound pressure level at 1 m	90 dB(A)

rating gives is how much electrical power can be fed into it without damaging it, or at least causing distortion. There is no automatic guarantee that the higher powered of two units will produce the better output, or even the *louder* output.

The *efficiency* of a loudspeaker is low – that is, the proportion of sound power coming out compared to the electrical power fed in is small. The efficiency of a high-grade monitoring speaker may be only 1–2%, so that for every (say) 100 W fed in only 1 or 2 W of acoustic power emerge, the remaining 98–99 W appearing as heat. Generally, the better the loudspeaker the lower its efficiency, so that a good quality loudspeaker rated at 100 W may actually produce no more noise than a less good one rated at 50 W.

Loudspeaker cables

Over recent years there have been available on the market some very expensive cables for connecting loudspeakers to amplifiers. It is pretty safe to say that properly conducted tests of the A–B type have shown them to have no advantages over conventional cables of adequate current rating. It is clear that a connecting cable must have a total resistance (i.e. in *both* conductors) which is much less than the impedance of the loudspeaker, otherwise power is wasted in the cable. The situation is complicated by the fact that the impedance of a loudspeaker can vary widely with frequency but is likely at some frequencies to be greater rather than less than the stated figure. Thus, an 8 Ω speaker will probably have an impedance of around 8 Ω over much of the range, but becoming greater over a narrow band. So, if the cable for this unit has a resistance of 1 Ω or less there should not be excessive loss in it.

As an example, a typical 13 A mains cable has a resistance of about 20 Ω/km in each conductor. Since there are two conductors, a 10 m run of this cable will have a total resistance of roughly 40/100 = 0.4 Ω, which is very small compared with the probable loudspeaker impedance. For most applications, cable rated at 10 A will be adequate.

If there are very long cable runs, it may be worthwhile using a system known as *100-volt working* (see Part 2 of Chapter 12).

Professional and domestic standards

Finally, it is important to make clear one important difference between professional and domestic equipment in regard to signal levels. The reader will have gathered from what has been said here that the standard level ('zero level') adopted in professional equipment is 0 dB (= 0.775 V). Almost all domestic, and indeed some semi-professional equipment, adopts −10 dB (0.25 V) as the reference. This is of no consequence unless there is a mixture of the two types of equipment, and even then any difficulties can be overcome. It is best, though, to be forewarned.

Questions

1. On a standard PPM, what meter reading is taken as the maximum signal level if distortion is to be avoided?
   a. 4   b. 6   c. 7   d. 8

2. What is the likely electrical efficiency of a high-grade monitoring loudspeaker?
   a. 0.1%   b. 0.5%   c. 1–2%   d. 10%   e. 50%

7 Stereo
Part 1

The word 'stereo' comes from a Greek word meaning 'solid'. 'Stereo-scopic vision' means perceiving things as being solid when viewed with two eyes. Although the term '3D' has become the standard (easier to spell!), 'stereoscopy' is still basically correct. From there it was but a small step to extend the 'stereo' part to sound and hence we have 'stereophony', or 'stereo' for short.

There is a sort of solidity, in a sense, when sound images are spread out in front of the listener. Here we shall deal with the usual form where there are two loudspeakers, two tracks on a tape, and so on. Actually, stereo sound could have any number of such *channels* – there are good reasons for saying the more the merrier – but practicalities such as economics mean that two is the norm, at least at the present. This is not to overlook the fact that there are various commercial versions which have three loudspeakers, often with the main bass speaker in the centre. This is justified on the grounds that very low frequencies are often somewhat indeterminate in their direction. However, we shall deal with two-channel systems.

In this chapter we shall consider how stereo works and look at techniques for producing stereo.

How stereo works

To begin with, we need to consider how we locate sounds in real life. A number of factors come into this, including visual clues. At first glance, one might think that the relative sound intensities at the two ears would be the major ingredient in sound location, but while this can have signif-icance in some conditions, the single most important factor turns out to be the *time of arrival difference* at the two ears.

If a sound source is directly in front of a person, then there is clearly no difference in the time that sounds from this source take to arrive at the ears. In the case of a sound arriving from any direction except the front, then there will be a time-of-arrival difference. (We will not concern ourselves here with sounds which originate from above or below the level of the ears.) In the extreme case when the sound source is right at the side, i.e. 90° from the front, the time-of-arrival difference is roughly 1 ms (1 millisecond = 1/1000 second). What is rather interesting is that under fairly good listening conditions most people can detect when a sound source has moved through about 1° from the front axis, and this represents a time-of-arrival difference at the ears of roughly 10 μs – *ten millionths* or *one hundred thousandth of a second*! How the brain does this is not clear.

Time of arrival difference – In general, the sound from a particular source will arrive at a person's ears at slightly different times. This is important in detecting the apparent direction of the source.

(The question of how we locate sounds which are above or behind the head is to some extent still a little obscure. Head movements help to pin down the source of a sound. Most people turn to try to look at the source of a sound and for each head movement the brain can carry out computations which enable the listener to narrow down the uncertainty. Also, there is evidence that the folds in the outer ear – the visible portion – may create multiple reflections at high frequencies and the pattern of these reflections will depend on the angle of incidence of the sound.)

It follows that to recreate directional information for a listener, at least in a frontal area, we need to create time-of-arrival differences at the person's ears.

Assuming that the listener will be sitting in front of a pair of loudspeakers, and equidistant from them, there can be no significant time-of-arrival differences at the person's ears. Can they be created in the studio? At first sight, this would seem to be the obvious thing to do and in fact this was what was done in the early days of commercial stereo by spacing the microphones some distance apart. Unfortunately, this does not usually work at all well. With a pair of microphones spaced, say, a couple of metres or more apart, the listener is apt to hear two separate sound images, one at each loudspeaker and nothing in between – a situation usually described as *hole-in-the-middle* or '*ping pong*' stereo. (Agreeable stereo can be produced if the microphones are not too widely spaced – perhaps no more than a metre.)

Now it so happens that if the electrical signals in the two channels – the left and right paths from microphones to loudspeakers – differ from each other *only in amplitude*, then a reasonable spread of sound images between the loudspeakers can be expected. The explanation for this is beyond the scope of this book, but there are readable explanations in some of the books listed at the end. At the moment, we will simply state that, for loudspeaker listening, there need to be amplitude differences between the two channels – *inter-channel differences* – and paradoxically, *no* timing differences, between the signals in them.

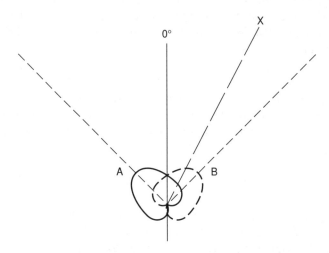

Figure 7.1 A coincident pair of cardioid microphones

Methods of producing inter-channel differences

Broadly there are two techniques, although in practice their usage often overlaps. These are:

1. Using a pair of directional microphones very close together. They are angled symmetrically about an imaginary line which points to the centre of the recording area. Figure 7.1 shows two cardioid microphones positioned in this way. A pair of microphones with the diaphragms as close together as possible form what is known as a *coincident pair.*

    If sounds arrive from directly in front of the two microphones – 0° in the diagram – then both microphones will receive the same sound levels and the outputs will be exactly equal. (It is very important that the two microphones are closely matched in performance. The same is also true of loudspeakers in stereo.)

    A sound arriving off the centre axis of the microphones, say from X in the diagram, will result in a larger output from the right-hand

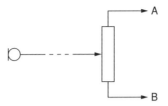

Figure 7.2 A panpot

microphone than from the left-hand one and this then gives us the required inter-channel difference.

2. Purely electrical methods. Figure 7.2 shows in simplified form what is termed a panpot (from 'panoramic potentiometer'). A potentiometer is a variable electrical resistance with three terminals. Two are at the ends of a resistive track, while the third is connected to a slider whose position along the track may be varied by means of a knob.

Coincident pair – Two directional microphones with their diaphragms mounted as close together as possible.

The position of the slider, which in this case is connected to the output of a microphone, determines what proportion of the microphone's output is fed to which channel. This then gives us a second way of producing inter-channel differences.

Terminology

It is more or less universal in the UK professional world to use the following conventions in stereo:

The LEFT channel is denoted by the letter A and, where appropriate, the colour RED is used to indicate this – for example, on cables and in diagrams.

The RIGHT channel is indicated by the letter B and the colour is GREEN.

This professional convention is easily remembered: in writing the alphabet A is on the LEFT of B; in navigation RED indicates port (LEFT) while GREEN means starboard or RIGHT.

Note, however, that on domestic equipment L and R are normally used for left and right, and also the colour convention is different: WHITE is commonly used for LEFT and RED for RIGHT. To add to the confusion, headphones may have YELLOW for left, with RED still for right.

PROFESSIONAL STEREO CONVENTIONS
A – left, colour Red
B – right, colour Green

Stereo listening

We have already implied that the listener should be on the centre line between the two loudspeakers and that these should be a 'matched' pair. And not only should they be matched in quality terms, but they must be 'balanced' so that if fed with exactly the same signal (i.e. mono) the resultant sound image is truly central. (How to check on the matching is explained below, as are ways of obtaining a mono signal.)

If the balance is incorrect it is then a matter of putting this right by either adjusting the gain (volume) controls on the amplifiers feeding the loudspeakers or adjusting the *balance control* if there is one, as there usually is on decent home hi-fi equipment.

The ideal listening position, then, is at the apex of an equilateral triangle formed by the listener and the loudspeakers, but in practice the listener can be further back from this point. How much further back depends on circumstances, mainly the shape, size and symmetry of the room.

The ideal listening room will be acoustically symmetrical between left and right. For example, it is undesirable to have a bare wall on one side and a window with heavy curtains over it on the other side. If the listener is not too far from the loudspeakers, some lack of symmetry may not be too serious as the direct sound from the loudspeakers will be much stronger at the ears than sound reflected from the walls. However, the further back the listener gets, the greater the influence of reflected sounds is likely to be. Few rooms can be truly symmetrical, so that there is a limit to the distance from the loudspeakers at which good stereo is heard.

The loudspeakers should not be too close to the walls. In many circumstances about half a metre from the back wall can be acceptable, and the space between them should not be reflective to sound. In a domestic situation, this can be difficult to achieve, although in the author's sitting room the space between the loudspeakers is filled by a fireplace designed deliberately to look like a traditional Cotswold stone wall. This seems to be perfectly satisfactory, possibly because there are sufficient irregularities to break up at least some of any reflected waves, and also the edges of the limestone pieces may be slightly absorbent to sound waves.

One notably unsatisfactory listening position is closer than the apex of the equilateral triangle, as images may become somewhat unreal. It can be a little like listening to stereo headphones.

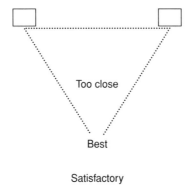

Figure 7.3 Listening positions

Having said that the listener should be on the centre line between the loudspeakers, a reasonable question to ask is, how far off this centre line can the listener be before stereo images become excessively left- or right-handed? The answer in brief is that for really critical listening there is very little latitude indeed, perhaps 2 or 3 cm, with a loudspeaker spacing of a couple of metres. However, there can be pleasing stereo listening, which is a different matter, when the listener is noticeably off the centre line. Much can depend on the particular loudspeakers. Some types will give a reasonable spread of sound when the listener is a metre or more off the centre line, assuming again a 2-metre base line.

A few words of reassurance about listening

Some of what has been written above could easily seem very worrying to the ordinary person who simply likes to listen to reasonable stereo. To such a person we would simply say, 'Don't worry!' Don't rearrange your sitting room to make it acoustically symmetrical; don't fill in a window or curtain a facing wall. If you have enjoyed your music up to now then go on enjoying it. (In a professional environment, though, there will have to be more care taken in designing listening areas.)

It might be, of course, that peculiar imaging has been a problem and there are hints here that might help. It's rather like the cases in my own experience when people with loudspeakers of which they were very proud had the loudspeaker responses measured. The almost inevitably uneven graphs caused great distress – really quite unnecessarily. These were cases of a little knowledge being, not dangerous, but disturbing – or perhaps one could say that mild ignorance was bliss!

Stereo loudspeaker matching

This means ensuring that the quality of output of the two loudspeakers is as near as possible the same. The best way of checking this is to feed both loudspeakers with mono music which contains a wide range of frequencies.

Assuming the loudspeakers are balanced in loudness then all frequencies should appear to come from mid-way between them. If, however, this is not the case and, to take an example, the high frequencies come from a point to the right of the centre, it means that there is excessive high frequency output from the right loudspeaker or, and much more likely, there is a high frequency deficiency from the left loudspeaker. A possibility in that case is that the tweeter of that loudspeaker is faulty.

We have said rather glibly that the easiest way of checking the matching, and also the balance, of the loudspeakers is to feed them with mono music. The trouble is that it may not always be easy to find a source of mono music! Professional equipment is usually provided with a switch which can convert stereo material into mono for purposes such as this. Home equipment often does not. The problem is not, however, insuperable. Some radio receivers intended for connection into a hi-fi system have a mono/stereo button to be used when the broadcast stereo is weak and noisy but the mono version is acceptable. This could be used to provide either mono music directly or a mono music tape.

Things become a little more difficult if there is no mono/stereo button. A partial solution, albeit without a wide frequency range, is to tune the receiver to a news broadcast when the newsreader will almost certainly be central and therefore mono.

Phase

This means, in essence, that the polarity of the wiring is the same in both channels of a stereo system. Figure 7.4 illustrates this.

In Figure 7.4(b) the electrical connections have, at some point, been reversed in ONE of the channels. The effect is that a sound wave compression is emitted from one loudspeaker when a rarefaction is emitted from the other. This is a very unnatural effect for a listener, being something that does not occur in real life, or if it does, only briefly and incompletely.

The result of out-of-phase stereo is that images are very difficult to locate. They may appear to come from behind or even inside the head and they are apt to move if there is any attempt to locate them by turning the head. It is not generally a pleasant experience and prolonged listening to out-of-phase stereo can induce a headache. There have even been reported instances of physical nausea.

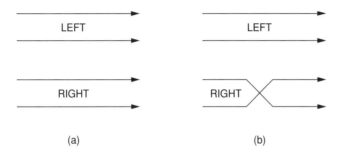

Figure 7.4 (a) Correct phasing   (b) Incorrect phasing

How obvious an out-of-phase condition is can depend very much on the type of programme material. With wide stereo images and little central sound an out-of-phase signal may not be at all obvious. It will be most apparent with mono or central images. This, then, is the way to check for out-of-phase connections. Play a central mono signal (and here a simple news bulletin will be quite adequate) and check that there is a clear central image. If there is any doubt then the wiring of one of the channels should be changed over, and probably the easiest way of doing this will be at the terminals of one of the loudspeakers. If the system is found to be out of phase, and the wiring to the loudspeakers is obviously correct, it will be necessary to check through the rest of the system. This will be an especially useful exercise if other equipment is likely to be added.

Questions

1. What is the approximate time-of-arrival difference at the two ears of a person when a sound source is 90° from the front?
   a. 10 μs   b. 0.01 ms   c. 0.1 ms   d. 1 ms

2. What is a likely cause of 'hole-in-the-middle' stereo?
   a. Out-of-phase conditions
   b. Microphones too widely spaced
   c. Signals in one channel being much larger than in the other channel
   d. Mismatched loudspeakers

3. A panpot is set with its control fully clockwise. Where is the image from it likely to be in the final stereo image?
   a. Fully right   b. Fully left   c. Half right   d. Half left

4. What should be a clear central image in a stereo scene is difficult to locate by listening. What is the likely cause?
   a. There is an out-of-phase condition somewhere
   b. Panpots are being used incorrectly
   c. A coincident pair microphone is mounted upside down
   d. One of the microphones is distorting

7 Stereo
Part 2

Microphone techniques for stereo

Here we can look a little more deeply into microphone usage for stereo. Various techniques exist, but here we shall look at three, and only briefly at two of those. The reader who wishes to go further into the matter is referred to the list of books headed 'Further reading'.

The first technique, and in many ways the simpler to understand, is that of *multi-miking* – making use of several microphones whose outputs are 'positioned' in the stereo image with panpots. From the point of view of the non-professional user, the difficulty with this approach may be not having a sufficient number of inputs and associated circuitry on the mixing equipment. Also, of course, a large number of microphones may be required – in the opinion of some professional sound balancers the drum kit alone may need six to ten or even more!

The main advantage of the multi-microphone approach is that there can be good control over the sound balance – that is, weak sections of, say, a band or choir can be raised in level to match louder sections.

The disadvantages are:

1. Technical complexity can be considerable and, paradoxically, can sometimes make it more difficult to achieve good results. There is also more to go wrong!
2. It is possible for microphones to form unwanted stereo pairs. This means that in addition to the panpotted images there may be spurious additional images caused by more than one microphone picking up the same sound. In practice, this is unlikely to mean that there will be multiple images; rather the panpotted images will be shifted from their intended positions. As a further hazard, moving a fader to alter any one microphone's output may result in some of the shifted images moving. The only way to avoid this effect is to have all the instruments, band sections or whatever 'closely miked', i.e. have the

microphones as close as possible to their respective sources. This can sometimes be difficult, especially as it is not always easy to retain the characteristic sound of an instrument with closely positioned microphones.

In short it's fair to say that multi-microphone techniques are generally best not attempted unless there are many microphones available and the mixing equipment is suitably comprehensive.

The second microphone approach is to use coincident pairs, which we have already referred to. Here a great deal depends on the polar diagrams of the two microphones. The assumption must be made that the diaphragms of the two microphones are as close together as possible, although it must be admitted that satisfactory results can be obtained when they are several centimetres apart (see 'Binaural stereo' on p. 101).

The main effect of the polar diagrams is on the *angle of acceptance* of the microphone pair. By 'angle of acceptance' we mean the angle in front of the microphones, which corresponds to a sound image stretching the full distance between the loudspeakers. Photographers will realize that the angle of view of a lens is a very similar concept.

Knowledge of the angles of acceptance for different polar diagram systems is essential if satisfactory results are to be obtained. However, it isn't difficult to carry this information in the head when it is remembered that hypercardioid microphones have polar patterns which are intermediate between figures-of-eight and cardioids. Then, for microphones which are at 90° to each other, the angles of acceptance are approximately:

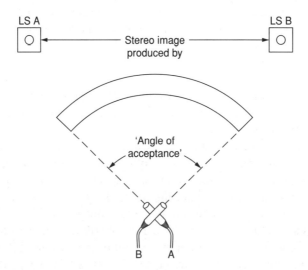

Figure 7.5   'Angle of acceptance' for a coincident pair

Figures-of-eight   90°
Hypercardioids   135°
Cardioids   180°
(135° is halfway between 90° and 180°)

Notice that we are assuming the microphones are at 90° to each other. This is fairly standard, and a few degrees difference on either side of 90° is unlikely to have any significant effect, although the angle can be increased considerably with cardioids, resulting in a wider stereo image.

Perhaps the big question now is, what polar diagram system to use when? Often, of course, there is little choice. Commonly the only microphones available are a pair of cardioids (as, for instance, in the built-in stereo microphones of most camcorders). However, the positioning of a pair of coincident microphones, of whatever polar pattern, is largely a matter of geometry.

Let's assume we have two cardioids – and we repeat that they should be as closely similar as possible, same manufacturer, same type number and so on. Figure 7.6 shows the probable image spreads for two positions for such a pair of microphones. In (a), the microphones would be close to the front of the sound source (band, choir, orchestra, etc.). In (b), the microphones are further back so that the axes of the cardioids are pointing approximately at the edges of the source. Remember that the angle of acceptance for a pair of cardioids is about 180°. In (a), the spread of sound images will fill the region between the loudspeakers almost completely;

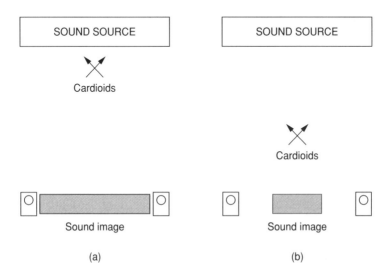

Figure 7.6 Likely image spreads for a pair of cardioids

in (b), only a part of the region will be filled. Note that the state of affairs in Figure 7.6(b) is not necessarily wrong. A solo piano will usually sound a bit silly if it appears to be the full width between the loudspeakers! Equally, a classical string quartet should normally fill only part of the picture.

With a rock band it is difficult to say what the reproduced image should be as there is often no attempt to reproduce the reality of the original sound.

In either case, though, reverberant sound will tend to fill the stereo image, thus adding to the realism, even if the basic image of the instrument(s) is narrow. It all depends on what will seem reasonable to the listener on the replay.

A pair of figure-of-eight microphones has rather different characteristics. To begin with, the angle of acceptance is smaller than for cardioids (90°) and for full loudspeaker-to-loudspeaker coverage the microphones should be further back – about where the two cardioids are in Figure 7.6(b).

Secondly, there is a rear pick-up region of 90°, which may or may not be a good thing. For example, it can be useful if plenty of reverberation is wanted; on the other hand, there may be unwanted noises from the rear – a coughing audience, for instance! A final and sometimes very important point is that the side quadrants are out-of-phase regions and this means that sounds arriving from these directions may be difficult for the listener to locate. The sides therefore have to be regarded as 'forbidden zones', although reverberation which arrives from the sides doesn't suffer any peculiar effects because it is, by its nature, so diffuse that phase has no real meaning. This is illustrated in Figure 7.7.

The point has been made that, other things being equal, figure-of-eight microphone pairs need to be further back than cardioids. There are some pros and cons as a result:

1. The further back the microphones, the better (probably) will be the relative balance between the front and back of an orchestra or band. Put another way, a close pair of microphones such as cardioids may result in the nearest instruments or voices being too loud in relation to the further ones. The relative sound perspectives (see Chapter 5) will be exaggerated.
2. If the acoustics of the hall or other recording area are poor, for example with excessive reverberation or external noise, then closer microphones like cardioids may help.
3. The closer the microphones are to the band, orchestra or choir, the more important it may be to have additional mono microphones to reinforce the more distant sections. Then panpots have to be used

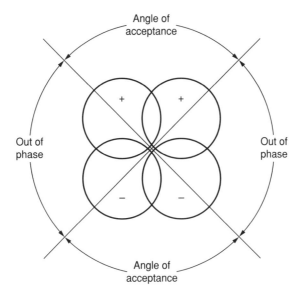

Figure 7.7 Figure-of-eight microphones as a coincident pair

to put the outputs of these microphones into their correct positions in the 'picture'.

It may be worth mentioning that, given a good band, orchestra or choir in favourable acoustics, a pair of figure-of-eight microphones can, in the opinion of many experienced sound balancers, give the most natural reproduction. It is also the simplest arrangement technically speaking.

A hypercardioid pair has not been mentioned but the reader will quickly be able to adapt what has been said about cardioids and figures-of-eight to hypercardioids, remembering that they can be regarded as being half way between the other two.

A third microphone technique is that of *binaural stereo*. It is intended primarily for headphone listening, but carefully done gives results which are acceptable on loudspeakers. The idea is that the ears of a person are replaced by microphones, or at least microphones are placed at, or even in, the ears. The theory (rather incorrect!) is that if those microphones are connected to headphones the listener is 'transported' to the position of the wearer of the microphones.

Special dummy heads have been designed and marketed commercially, but much simpler and cheaper is to have a disc of perspex, some 6–8 mm thick and 200–250 mm diameter. Two small microphones are mounted on opposite sides of the disc, on its axis and roughly 100 mm from it. These form the stereo pair. (Note that the two microphones have a spacing which is approximately that of human ears.) Interesting results can be obtained, although most headphone listeners find that the sound images tend to be behind them. There is scope for experimentation by the reader with two small microphones and modest workshop facilities.

Headphones for stereo monitoring

Having mentioned the use of headphones, it may be worth adding a comment about their use in stereo.

To begin with, serious monitoring of stereo is almost impossible with headphones, because the positions of the sound images in general bear almost no relation to those perceived when listening to the same stereo material on loudspeakers. Some sounds can appear to be behind or even inside the head. I once carried out some experiments to see whether there was any correlation between the positions of 'headphone images' and 'loudspeaker images', but even experienced listeners differed so much among themselves in their location of headphone images that the final conclusion was there was no significant correlation at all.

Lastly, in setting out to do a stereo recording, it is worth studying the situation in advance to form some idea of what microphones to use and where to place them. Then, and this is most important, a rehearsal should be used to test as thoroughly as possible whether the optimum microphone positions have been achieved. This means listening critically, preferably on loudspeakers. If this cannot be done in the recording venue, then it means making a recording and playing it back in the best possible listening conditions.

All this is often easier said than done, especially as there are complications like the fact that the presence of an audience will almost certainly modify the acoustics, and the musicians will play louder and better when there is someone to play to! However, with experience it is possible to make some allowance for these things.

Questions

With a coincident stereo pair of microphones angled at 90° to each other, what are the effective 'angles of acceptance' for the following?

1. Cardioids
   a. 90°　　　b. 135°　　　c. 180°

2. Hypercardioids
   a. 90°　　　b. 135°　　　c. 180°

3. Figure-of-eight
   a. 90°   b. 135°   c. 180°

4. Omnis
   a. 90°   b. 135°   c. 180°

5. With what coincident pair arrangement is there likely to be a significant out-of-phase region at the sides?
   a. Cardioids   b. Hypercardioids   c. Figures-of-eight   d. Omnis

8 Sound mixers
Part 1

By the term 'sound mixer' we mean equipment designed to accept the electrical outputs from microphones and other sources, such as tape machines and CD players, and allow an operator to combine these in ways which are artistically and technically satisfactory. Mixers provide some or all of the following features:

1. Amplification of very weak signals such as those from microphones.
2. Control of the *level* – i. e. signal voltage – of the various sources.
3. Mixing of the various sources in whatever proportion is desired.
4. Monitoring by means of meters, loudspeakers and headphones (the loudspeakers will not generally be built-in).
5. Electrical manipulation of the signals by adjustment of their frequency responses (a process known as *equalization* or EQ; see below) or the addition of artificial 'echo' or other effects.
6. Control of the level of the output signal so that, for example, recordings are not distorted because of signals of too high a level.
7. On large installations certainly, and possibly on small equipment, facilities for providing communications. At its simplest this means letting a presenter hear the programme passing through the mixer with provision to hear, at the same time, instructions from a producer.

In very large broadcast installations, both for radio and television, and in recording studios there may – almost certainly will – be much more, but we will assume here that the interest is in fairly modest equipment (although if you have a fair understanding of what we're calling 'fairly modest equipment' you're well on the way to being able to understand big mixers). However, before we go further let us establish a few definitions, or perhaps more accurately, explain some of the terms we shall use.

Terminology

Signal – The electrical voltage which is the output of a microphone, tape machine, CD player or any other source.

Fader – A control which allows the level (i.e. voltage) of a signal to be varied. The volume control on a hi-fi set is essentially a fader. Practical faders may be rotary controls or take the form of sliders.

Channel – This means the route and the associated controls of a particular signal inside the mixer until it is combined with other signals.

Equalization (EQ) – Put very simply for the moment this means using tone controls, but as we shall see later the EQ controls on all but the very simplest mixers are likely to be much more comprehensive than on normal domestic equipment.

Figure 8.1 illustrates in simple form the terminology we have introduced above. Note that in this sort of diagram a straight line connecting blocks symbolizes an electrical path and may thus represent a *pair* of wires.

It may be appropriate here to mention that full-size studio mixers may well have 80–100 (or even more) channels and that the total number of controls can add up to a few thousand! We'll deal here with much simpler devices where the number of channels may be no more than four, with a maximum of perhaps 20 or 30.

Let us now begin to look in a little more detail at the parts that have been outlined above.

The basic channel

It may be helpful to refer to Figure 8.2 from time to time while reading this chapter. (The diagram is repeated so that you don't have to keep turning pages to look at it.) It is a *block diagram*; that is, it shows the relationship and connections between sections of the equipment but it does not give details of the circuitry.

There are a few points about Figure 8.2 which probably need clarification:

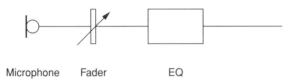

Microphone   Fader   EQ

Figure 8.1 A much simplified section of a channel

1. The block marked 'dB' is simply a device which reduces the voltage of a high level signal to make it compatible with the microphone level input. (The term 'high level' is explained below.)
2. An arrow through a symbol means that the circuit represented is adjustable by some sort of control knob.
3. What is termed a 'variable gain amp' is an amplifier with a 'volume control'. It constitutes the *channel sensitivity control* referred to later on p. 108.
4. The blocks marked ϕ and EQ have a switch below them. The idea of this is that the unit is bypassed, or switched out, when the switch is closed, thus making that unit inoperative.
5. An *insert point* allows other items of equipment to be connected into the channel. The two × symbols represent sockets. Insert points are not very likely to be found on the smallest mixers.
6. Not shown in the diagram, but important features of all but the smallest mixers, are 'auxiliary outputs'. These are used for feeding such things as artificial reverberation devices and loudspeakers used for audiences. (The term 'Public Address' (PA), although not terribly accurate, is used for this.) Auxiliary outputs may be connected to the channel before the fader, or after it, or switchable from 'pre-fader' to 'post-fader'. There's also a facility called 'Foldback', FB for short. This and PA will be explained further in Part 2 of this chapter.
7. A further point is that a practical mixer will have a number of amplifiers in each channel (and elsewhere). These are not shown in the diagram for the sake of simplicity.

How do we connect the sources to the mixer? In the case of a non-permanent set-up the microphones and other sources will generally be plugged into the back of the mixer. This can be the cause of some confusion depending on what types of plugs and sockets are used. We go into

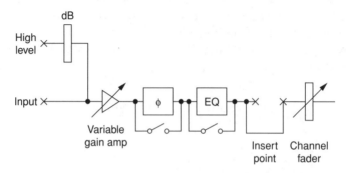

Figure 8.2 A basic channel

this in a little more detail in Part 2 of the chapter. However, it is almost invariably the case that any mixer will have to cater for two types of source:

- 'low-level', sometimes called 'mic-level', sources; and
- 'high-level', sometimes called 'line-level', sources.

The difference between the two is a matter of the voltage output of the source. Low-level (mic-level) signals are just that – the sort of voltage which a typical microphone can deliver, and this is likely to be of the order of a millivolt (1/1000 V). High-level, or line-level, sources include tape and disc machines, and their outputs are probably going to approach 1 V, possibly more.

Expressed in decibels, low-level signals are typically in the range of –70 to –50 dB, relative to zero level (see Chapter 6, Part 2); high-level signals will be in the range from –20 to +10 dB.

It is important to make sure that a source is fed into the correct level input, not that any damage is likely to occur, but if the very small voltages from a microphone are fed into a high-level input the resulting output will be exceedingly weak. A high-level source fed into a low-level input will, almost without doubt, become very badly distorted.

To get round this problem, some mixers have two separate inputs for each channel, one for high level, one for low. On others there may be only one input for each channel but a switch (maybe a push-button) is used to select the appropriate level. Figure 8.2 shows two separate inputs. The block marked 'dB' represents what is known as an *attenuator*. It is essentially a system of resistances and its function is to reduce the voltage of the high-level signals to be comparable with those from, say, microphones.

Important features of any mixer

Channel sensitivity (or 'gain')

It is often necessary to make adjustments to the signal level from a source in addition to the high/low-level selection we have just mentioned. Different microphones may have different sensitivities, or compensation may have to be made for the loudness of the source. This could be handled with the fader, but it is often not convenient to do this for a reason which will be explained under the section headed 'The fader'. A channel sensitivity control or 'gain' control is generally a small knob and may sometimes be concentric with a rotary fader, or it may be near the top of the mixer, furthest from the operator. (It's normal to position the controls which are used least furthest from the operator. This includes channel sensitivity controls. Faders, however, are used more often than anything else and are thus placed nearest to the operator.)

Phase reverse (symbol φ)

We have already mentioned the importance of correct phasing in the chapter on stereo. It can be important in mono as well. Imagine two similar microphones equidistant from a person speaking. If both are faded up then the combined outputs will cause an increase in level *provided that they are wired correctly*. If for some reason they are not, so that fading up both results in a cancellation effect (never in practice 100%), the combined signal is low in level and often sounds very 'thin' and 'toppy'. The solution is to bring about a phase reverse somewhere, and a suitable switch or button can do this.

One argument may be that all equipment should be wired correctly so the problem should never arise. True, but just occasionally mistakes do occur. In practice, the most likely source of trouble is in the cables – more specifically in the connectors at the ends of the cables. It is all too easy for someone to make these up for a specific purpose and fail to get the phasing right. Also, some microphones do not conform to the European convention. Never assume a perfect world when it comes to wiring standards!

A further use for phase reverse switches, although not perhaps in channels, is to check that all is well during a rehearsal for a stereo recording. It is sometimes difficult to be sure that the phase as listened to on a loudspeaker is correct. A phase reversal in the feed to one loudspeaker can set the operator's mind at rest – or not!

Channel fader

This is the most important single control on the channel. It can be thought of as a volume control, as we have suggested, but on a proper mixer it will be much more accurately calibrated than is likely to be the case with a domestic hi-fi system. On small mixers the channel faders may be rotary controls. On large mixing consoles linear faders are used as these take up less space sideways, so that, say, twenty linear faders can be fitted into a unit 500 mm wide. A big advantage with linear faders is that several can be controlled simultaneously with one hand. They must, however, have a reasonably long 'travel' to allow accurate operation. The miniature sliders found on some domestic and semi-professional equipment are poor in this respect.

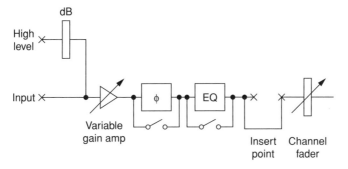

Figure 8.2 repeated

Fader calibration is likely to be reasonably accurate on all but the cheapest mixers. Some will have an arbitrary scale. Others, and these are to be preferred, will have a scale marked in decibels. In these a typical scale will be marked with numbers going from −∞ (minus infinity = fully faded out) through 0 to +10. The 0 position represents the setting at which, if all is correctly adjusted, the output from a professional mixer will be at about our old friend zero level. In other words, the 0 position is the normal working position, but there are several decibels 'in hand', meaning that the fader can be moved to appreciably higher settings if necessary. This matter of a 'normal working position' on a fader explains the reason for a channel sensitivity control. Any source can be adjusted until the mixer's output is at about zero level when the fader is set to 0 (or maybe 10). All stages of the electronics should then be operating so as to give optimum performance.

The scale on the fader, if it is marked in decibels, should be even, so that equal movements of the fader produce equal changes in the level in decibels. (Remember that equal changes in decibels result in roughly equal changes in loudness as judged by the ear.) At the bottom end of the scale, as −∞ is approached, the decibel increments can be much greater, as now the signal will be very low and a fade-out to (or a fade-up from) silence is what is wanted.

Equalization (EQ)

Almost any mixer, however modest, will have some EQ even if this is no more than a simple bass-cut control in each channel. On big and expensive mixers the range of EQ facilities may be very impressive and versatile. Here we shall deal with the basic types of EQ, the sort that might be found, not perhaps on the cheapest devices, but on medium cost units.

Briefly, the purpose of EQ is to modify the frequency response of a sound signal in order to:

1. 'Clean up' the signal. That is, to remove or at least reduce the worst effects of unwanted sounds such as traffic noise, rumble caused by wind blowing across the microphone, sibilance (excessive 'sss' sounds in speech), boominess from reverberation, etc. The list is almost endless.
2. Enhance or improve the clarity of the wanted sound by, for example, boosting certain parts of the frequency range.
3. Produce special effects. A good example is the simulation of telephone-quality speech in drama.

Before looking at the fundamental EQ controls, a word or two of clarification about the terminology may be helpful. Anyone with the slightest knowledge of music knows the terms 'bass' and 'treble', and these words are also used in relation to EQ controls. The difference in usage comes in the fact that, taking 'treble' as an example, instead of confining its use to the upper registers of voices or instruments, with an upper frequency of perhaps a few kilohertz, a treble EQ control could extend to 15 or more kilohertz. The terms as used in EQ are therefore much wider in their range than a musician might expect.

1. Bass cut or lift. This reduces or raises the bass content of the signal. Usually the control knob reduces the bass when turned anti-clockwise and increases when turned clockwise. Many good hi-fi units have a similar facility. Judicious use of bass cut can be valuable in, for instance, reducing traffic rumble and boominess caused by excessive reverberation in a hall.

   Bass lift is much less useful but may on occasion be helpful when the sound quality seems 'thin', or perhaps the bass section of a choir could be improved with a little emphasis.
2. Top (treble) lift/cut. Top cut can be very helpful in reducing any hiss or sibilance. Top lift, like bass lift, is less useful but may go some way towards correcting 'woolly' sounds, or perhaps compensating for an incorrectly placed microphone.
3. Usually only found on the more expensive mixers is a *presence* control. This provides lift or cut to intermediate frequencies. The term originated some time in the 1960s, when it was found that raising a limited frequency band – around 2–3 kHz – by only a few decibels had the effect of making certain sources, vocalists in particular, sound nearer. (No one seems to have a very satisfactory explanation of why!) Lift can improve intelligibility of speech or give a little more impact

to vocals. Cut can sometimes make unwanted mid-range sounds less conspicuous.

4. Bass and treble filters. These are like the cut controls mentioned above in 1 and 2, but they have a much more drastic action. A deep bass rumble may be almost eliminated by use of a bass filter.

More information about EQ controls is given in Part 2. What we must emphasize in the very strongest terms is that the decision to use any EQ should be made only after listening to its effect. It is all too easy to affect adversely a wanted sound when trying to reduce an unwanted one. Experience helps, of course, to shorten the amount of time spent in trials, and also some knowledge of what is going on can be valuable.

An example of this is trying to get rid of the buzz caused by having microphones and cables too close to fluorescent lights. This noise contains a very wide range of frequencies, from 50 or 100 Hz right up to several kilohertz, and it should be obvious that any of the EQ controls mentioned above are going, at best, to affect only a small part of the buzz's range. In other words, with that kind of unwanted noise, EQ is of little use.

Output stage

Then we come to the output section of the mixer represented in Figure 8.3. The lines coming in from the left represent the output of different channels. The meter on the right is labelled 'PPM' but it may of course be some other kind of meter, and in addition to the loudspeaker there may be, indeed almost certainly will be, a socket for headphones. The 'fader' before the loudspeaker is, in this case, a simple volume control.

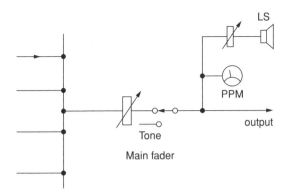

Figure 8.3 An output section

It should, of course, not be forgotten that in a stereo mixer all the components shown will be duplicated. We will explain the 'tone' switch in Part 2.

A famous and very highly respected manufacturer of professional mixers once produced a unit for mobile use in which the PPM was connected in error between the loudspeaker and its volume control. Quietening the loudspeaker also reduced the PPM reading. We leave it the reader to deduce why this was a mistake!

Questions

1. What is the main purpose of a channel sensitivity control?
   a. To act as an additional fader
   b. To reduce distortion
   c. To allow the channel fader to be set to an optimum position
   d. To allow the channel to handle weak signals

2. The symbol ϕ on a mixer indicates
   a. The mains on/off switch   b. Phase reverse control
   c. Phantom power switch   d. Bass cut

8 Sound mixers
Part 2

Inputs and connections

In dealing with the inputs to a mixer, or indeed any other connections, we must make clear a most important distinction which exists between professional and non-professional sound equipment: professional equipment invariably makes use of what is termed *balanced wiring* for connections between different items. Domestic equipment, on the other hand, almost always uses *unbalanced* wiring. Semi-professional systems may use either or both.

In balanced wiring there are three conductors: two carry the audio signal and one is for an earth connection, the latter often being in the form of a 'screen' of braiding. Care is taken to ensure that the two signal wires are as similar as possible to each other in electrical terms, which also means that they are geometrically symmetrical. The point about this arrangement is that if there is a nearby source of electrical interference, such as a mains cable carrying a substantial current, first the screen tends to 'protect' the signal wires. Secondly, if any interfering voltages are generated in the signal wires then, because of their close similarity in electrical characteristics, these voltages will be equal and will then tend to 'cancel out'. Some microphone cables are made up into what is known as 'star quad' form. Here there are four conductors, twisted around each other and then enclosed in a screen. Opposite pairs of the four conductors are connected together and the result is a cable which is highly immune to the effects of interference.

With unbalanced wiring, where there is, for example, a central conductor surrounded by a braid screen, the 'cancellation' referred to above cannot take place, although the screen gives some protection. However, such cables can be perfectly satisfactory for short lengths, such as in the links between a domestic cassette player and an amplifier, when 15–20 cm of connecting wire is all that is needed. Furthermore, the audio signal

Figure 8.4 Pin arrangements on XLR plugs and sockets

voltages in such instances are much greater than those in a micro-phone cable and are therefore likely to exceed greatly any interfering signals.

So, at the input to a mixer we shall hope to see balanced sockets. These will most likely be what are known as *XLR sockets*. Figure 8.4 shows the pin/socket arrangement when the connector is viewed end-on. Plugs have pins which are recessed into the housing; sockets have three holes, the ends of which are flush with the front of the housing.

A suitable microphone will therefore have an XLR socket (male) incor-porated in it; the cable will have an XLR female plug at the microphone end and a male plug at the mixer end, while the mixer inputs will be female. The screen connection at the microphone end will continue so that the microphone case is also a screen. (Note that we have used the terms 'plug' and 'socket' somewhat arbitrarily here, implying that a connector is a plug if it is on a cable and a socket if it is not.)

The terms 'male' and 'female' can, in some cases, be very confusing. Basically if there are pins it is male. Sometimes the distinction can be far from clear!

Another important type of connector is the so-called PO (= Post Office) jack. Although devised about a century ago, the design was so good that PO jacks have survived with little modification to the present day. There are variants – some for mono and some for stereo. Basic mono ver-sions are shown in Figures 8.5 and 8.6. Socket connections are shown in Figure 8.7.

Proper stereo jacks have a second conducting band and the sockets are modified accordingly.

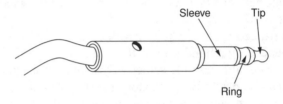

Figure 8.5 A PO ('Post Office') jack

Figure 8.6 Mono PO jack wiring

The British Post Office had responsibility for UK telephone services up to about 1981. After that, British Telecommunications (BT) took over, but things like the jacks referred to here at still known as 'PO' or 'Post Office' jacks.

We should emphasize that there are significant – but not always obvious – differences between the true PO jacks and the domestic versions. For example, the former are tapered, the latter are not, so that there is an inherent incompatibility between the two. Any attempt to use one type with the other may result in physical damage at the worst, and poor contacts at best. (Information about wiring conventions is given in the 'Data' section near the end of the book.)

An important practical point about PO jacks is that they must be clean. They are very reliable – otherwise they wouldn't have been in professional use for so long – but they are made of brass and this is a metal that tarnishes easily. Any kind of tarnish on a metal is likely to have a rectifying action – which means that there is a tendency for an alternating current to become unidirectional and action of this kind can easily convert any tarnished contact into a crude radio receiver. As a

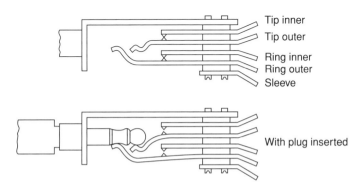

Figure 8.7 PO socket wiring

result, unwanted radio pick-up from a high power medium or long wave transmitter in the vicinity – up to 15 miles or more – can produce quite good but unwanted reception of radio programmes in a recording.

Consequently, an occasional cleaning session with any good propri- etary brass cleaner is very strongly recommended. The semi-professional type, with a chrome surface, isn't liable to tarnish.

Quite a lot of useful information about types of plug and socket can be gained from the study of professional and semi-professional catalogues, the latter being readily available at high street bookshops.

Equalization

Some readers may prefer to see the graphical representation of the types of EQ outlined in Part 1 of the chapter.

Bass lift/cut

This is also known as l.f. (= low frequency) lift/cut. Figure 8.8 shows typical curves.

In Figure 8.8, the maximum rate of lift or cut is given as '6 dB/octave'. This means that on the steepest part of the curve, when the control is set to maximum lift or cut, each doubling or halving of the frequency (i.e. a change of one octave) changes the level by 6 dB. Perhaps we should try to make this a little more meaningful with an example. Let's suppose that we have bass cut selected and the frequency below which it is going to be effective is 120 Hz. Now suppose that there is an annoying 50 Hz mains hum. Can we reduce it with the bass cut?

One octave below 120 Hz is 60 Hz and the level will then be 6 dB lower. Without going into detailed calculations, it can be seen that the 50 Hz hum will be reduced by only a little more than 6 dB. We said

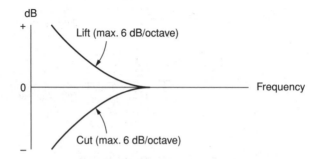

Figure 8.8 Bass lift and cut

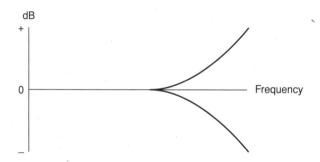

Figure 8.9 Top lift and cut

in Chapter 1 that a reduction of 10 dB is subjectively equivalent to roughly halving the loudness, so in this case the bass cut will probably do little to reduce the effect of the hum. Worse, it may adversely affect the character of the speech or music. A bass cut control may be very useful in other ways, but what we are saying is that 6 dB/octave is not necessarily going to be useful in trying to eliminate certain unwanted noises.

Top lift/cut

This is also known as treble or h.f. (= high frequency) lift/cut, and is illustrated in Figure 8.9.

Again, the likely maximum slope will be 6 dB/octave.

Presence lift/cut

This is sometimes called m.f. (= mid-frequency) presence. Figure 8.10 suggests that the maximum amount of lift or cut is about 15 dB. This

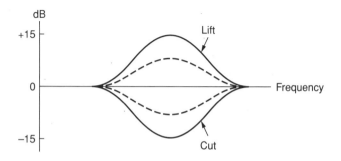

Figure 8.10   Presence lift/cut

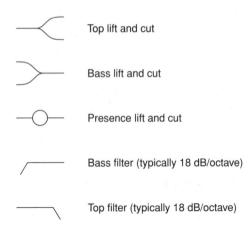

Top lift and cut

Bass lift and cut

Presence lift and cut

Bass filter (typically 18 dB/octave)

Top filter (typically 18 dB/octave)

Figure 8.11   Symbols used for EQ controls

may not always be the case – the manufacturer's manual should always be consulted for information of this kind.

The symbols used on mixers have become more or less standard, with usually only minor variations from one manufacturer to another. Figure 8.11 shows the commonly used ones.

Filters

We probably don't need to have graphs but we can state that the slope of bass and top filters is typically around 18 dB/octave and if we apply this kind of slope to the hum problem above we can see that a steep filter is likely to be much more use. At the same time, of course, there is a risk of deterioration of wanted material, so care and compromise have to be used. Filters, except on the biggest and most expensive mixers, usually have fixed characteristics and are simply switched in or out.

Other audio filter devices include:

• *Graphic Equalizers*. These consist of a largish number of separate controls, each one varying the level of a band of frequencies. The range of level is often from –15 to +15 dB. On smaller units there are eight such controls, each one covering an octave. Bigger ones might have thirty-two controls, each handling a third of an octave. The controls have slide knobs on the front of the unit so that they give the appearance of a frequency response graph – hence the name 'graphic equalizers'.

   They are almost always add-on units. That is, they aren't normally built into a mixer. Instead, they are plugged into insert points.

- *Parametric Equalizers*. I've never been sure how they got their name! They don't equalize parameters. They're basically filters which act over a narrow band, like each part of a graphic equalizer, but the band is continuously variable in frequency, width and height (or cut). One very good instance of their use is to remove an unwanted steady note, such as line-up tone which is being picked up from somewhere. The equalizer is set to lift and then the frequency is swept steadily until the unwanted frequency is lifted. At that point the lift is changed to cut, and with any luck the interference is drastically reduced without, hopefully, serious effects on the rest of the programme sound.

Public Address and Foldback (PA and FB)

These are both loudspeaker feeds taken from some point in the mixer, such as one of the channels or, more commonly, from a group of channels. The basic difference between them is as follows:

- PA is for the benefit of a studio audience. A good example is in a television sitcom, where an audience can laugh at appropriate places, this being picked up by microphones and becoming part of the show. (Canned laughter doesn't ever sound convincing, in my view!) Also, the presence of an audience can encourage the performers – there's nothing like a live audience to bring out the best in any actor, comedian or musician. Now obviously this is only going to work if the audience can hear the dialogue, but to preserve some appearance of reality the actors in a sitcom will generally have to speak in normal voices – and they won't be heard by much or all of the audience unless the audience has a loudspeaker feed of the dialogue.
- FB, on the other hand, is for the benefit of the performers. For instance, and again to take a fairly typical television situation, although this can apply equally to radio, there is a largish band in the studio. It's important for the different musicians to hear other sections of the band, and the easiest way of doing this is to let each section or individual have a foldback feed of the sections that are most important to them. If it's visually acceptable, then headphones can be used for the foldback. Alternatively, discreetly placed loudspeakers can be used. In a television play the foldback of, say, important sound effects must be from loudspeakers.

The 'Tone' switch in Figure 8.3

There is no point in careful monitoring at the mixer if, say, a tape recorder connected to the mixer's output is set to the wrong record level. A very useful – one might say, essential – facility on any good mixer is the ability to send a known test signal to the recorder first. In the professional and semi-professional world, this signal is known as 'line-up tone' or 'tone' for short. It consists of a sine-wave signal having a frequency of about 1000 Hz (the exact frequency is not critical) and an accurately defined level, or voltage. This is usually zero level, referred to in Chapter 6, Part 2, where it is stated that this corresponds to 0.775 V. In the mixer an internal circuit generates the tone and when the switch shown in Figure 8.3 is operated then the first thing is that the meter will indicate zero level ('4' if it is a PPM) and the loudspeaker will emit a 1 kHz note.

Assuming there is some kind of meter on the recorder, this should show that the record level control is correctly set. (It may be necessary to record several seconds of the tone and then play this back to check that the machine has been set correctly.) It is standard practice on professional tape machines to record a length of tone on the tape and then follow this by a few seconds of silence before the recording proper. This allows a check on replay settings no matter on what machine the tape is replayed.

Figure 8.12 A small professional sound mixer (courtesy Audio Developments Ltd)

Talkback

This is the simplest of communication systems. Many mixers, even quite small ones, incorporate a microphone. Operation of a switch somewhere on the unit connects the output of this to a presenter or performer(s). Typically a presenter, with his/her microphone, will hear on headphones the mixer's output, but talkback replaces this when the switch is operated. Instructions could be of the form of 'Go ahead in 10 seconds'/'when the curtain falls/rises' – the variations are endless.

In a studio the talkback is often via a loudspeaker which may be arranged to be muted during actual recording or transmission, headphones then being the performer's means of hearing the talkback.

The mixer shown in Figure 8.12 is a small, professional unit which is easily portable. It has a carrying handle which can be used, as in the photograph, to raise the unit to a more convenient angle. There are six channels, with slide faders and two 'group faders' on the right.

A fairly common arrangement on small mixers of this sort is to have two group faders. Each channel can be routed to either of them, so that

Symbol	Meaning
	Microphone. Based on a design which came out probably in the late 1940s. It had a spherical body and a flat gauze disk and was known as the 'apple and biscuit'
	Loudspeaker
	Wires joining
	Switch
	Amplifier. The triangle points in the direction of the signal path. There may or there may not be a box around the triangle
dB	*Fixed attenuator (or pad).* This reduces the signal level. The number of dBs attenuation may be stated
	Resistor
Variable circuit components are shown by an arrow passing through the symbol	
	Variable gain amplifier. The gain may be switchable in steps of a certain number of decibels, or the control may be continuous
or (traditional)	*Fader.* The BS approved symbol shows an idealized drawing of a type of fader which went out of fashion years ago. A modern fader is essentially a variable resistor

Table 8.1   Some symbols used in diagrams

they then become a stereo fader, or they can be independent mono faders.

There are two PPMs in the mixer shown. On small mixers there are often options: they can be chosen to operate as level indicators for stereo – one for left and the other for right, or possibly one for the M signal and one indicating S. Sometimes it is useful to have one for the main output and the other showing the level of some other output or input.

Mixers of this kind are designed to be flexible so that they can readily be used in a wide variety of situations.

Questions

1. You are given the job of laying out microphone cables, via a fairly complicated route, from a studio area to where the mixer will be placed. At this time, neither the microphones nor the mixer have been provided. Which way will you have the cable?
   a. With XLR sockets in the studio area and the plugs in the mixer area?
   b. With the XLR plugs in the studio area and the sockets in the mixer area?

2. Which set of figures is correct about line-up tone?
   a. Zero level, 100 Hz
   b. +3 dB, 1 kHz
   c. About 0.775 V, accurate 1 kHz
   d. Accurate 0.775 V, about 1 kHz

9 Controlling levels

We have earlier emphasized the necessity of not allowing programme (signal) levels to exceed certain limits, and we have shown that the right type of meter will indicate when maximum levels are being approached. What we have not done is to explain how to avoid undesirable excursions of the signal, either into the noise regions by being too low or into distortion by being too high. What is more, the process of keeping the signal within limits must be done with the listeners' requirements in mind. In some cases, the prime need will be to maintain what we might term 'artistry' with the preservation of at least an illusion of a good range from loud to quiet, in others there may be a different requirement. Cassettes to be listened to in cars probably ought to avoid very quiet passages which would not be heard above wind and engine noise. (Equally, classical CDs with a wide dynamic range are often a waste of time in cars – unless the engine is abnormally quiet and there is almost no wind noise!)

In some professional circles, the term 'control' is used to mean the setting of the level – control of the signal voltage, if you like. We will use the same term here.

Broadly, there are two ways of controlling the level – it can be done manually or it can be achieved automatically. Both have advantages and disadvantages. We will take them in turn.

Manual control of levels

As it suggests, this means varying the level with a fader, and in those bare terms it sounds extremely easy. Sometimes it can be, but there are times when considerable skill is needed. It all depends on how much and in what way the programme signal is varying.

Take a straightforward recording of speech. A good speaker will maintain a reasonably constant loudness, changing enough to give interest to what is being said, but nevertheless keeping the level within a range of perhaps about 10 dB. (Remember that this could allow the loud parts to be about twice as loud as the quiet parts.) The fader could then be set so that the loudest parts of the speech are a little below the maximum permitted level as indicated on the meter. The quietest parts should then still be well above the noise levels. The reader might ask how to set about achieving this. At the rehearsal is the short answer. Even an experienced reader will probably be glad to have a chance to read through the script and, provided this is done aloud and in the way which will be used for the recording, it is an excellent chance to set the recording levels on the faders.

Alternatively, when the speaker and recordist have worked together before and know each other's methods, it may be sufficient to adopt the professional's time-honoured method and ask for 'A few words for level, please'. A typical sentence may well be enough.

The less experienced or nervous speaker can be a problem. Here one must expect changes in level, caused not only by variations in voice production, but also by changes in distance from the microphone. A typical 'working distance', that is the distance from source – in this case the mouth – to the microphone diaphragm, could be 50 cm. The speaker might unconsciously move closer by only 15 cm, and then get further back, again by only 15 cm, so the working distance has varied from 35 to 65 cm. This alone would result in about a 5 dB change in level – and a movement of 15 cm towards the microphone isn't a vast amount. Leaning forward a little to look more carefully at a script could account for it!

Compensating for level changes of this sort by operating a fader when the person's movements are quite unpredictable is almost impossible. It's possible even to get totally out of step, so that the fader is brought up when the person is nearer the microphone! This might sound ridiculous, but it's not all that rare.

Even worse is the case of an animated discussion between two or more people. As they begin to forget the microphone(s) it is very easy for the members in a small discussion group to sit back in their chairs, lean forward to make a particular point, or turn away to address someone on the same side of the table. In different circumstances this might be termed 'body language'. The sound recordist has other phrases!

Manual control here is again very difficult. Almost the only thing to do is to set the fader(s) to what one hopes will avoid overloads and then keep one's fingers crossed. Automatic level control, which we will deal with a little later, may be the answer.

It will be noted that the discussion group we have been using as an example is, by its very nature, impromptu and unscripted. Where there is a clear script things become much easier. In recording a play, for example, it should be possible to mark a script during a rehearsal or read-through to indicate quiet and loud parts. These can be anticipated and there can be judicious use of the fader to keep everything within limits – at least technically. To do so and at the same time preserve an artistically satisfactory effect calls for extra skill.

Music and drama both rely upon, amongst many other things, 'dynamics', which we can define as changes in loudness. These may sometimes be considerable but are very important in the performance. Music, especially classical music, played at the same loudness is very boring. An orchestra playing *fff* (molto fortissimo) may, at a distance of only a few metres, produce sound levels of well over 100 dB(A). A short time later the same orchestra may be at a pianissimo (very quiet) passage and producing no more than around 30 dB(A). This is a range of some 70 dB, and it could be much more. This is far beyond the capabilities of any analogue tape system. Yet, if the orchestra's sound level were evened out to a range of, shall we say 30 dB, the music would sound terribly uninteresting.

Dynamic range – the difference in dB between the loudest and quietest signal voltages.

This is where the skilled sound operator can make use of a particular characteristic of the human ear/brain system. We are much more aware of *changes* in loudness than we are of actual loudnesses in absolute terms. So, when a loud passage in the music is impending the fader can be taken down very slowly, over a minute or so if possible. Provided the drop in level is not too great, the listener will probably not be aware of this. Then comes the sudden loud passage, and if the preceding operation has been carried out correctly, the upper limit will not be exceeded. Similarly, if a quiet section is approaching, the level can be brought up gradually beforehand so that it is not lost in tape hiss.

Figure 9.1 shows the idea, albeit in a rather idealized situation. The point is that the *changes* in signal level – the dynamics – are largely maintained. If the time scale is long enough, then the slow changes brought about by the operator's use of the fader will probably go unnoticed.

This is a very skilled and tricky operation. It means that signal levels need to be marked in the score or script before the recording, so in the case of music it helps if you can follow a score. (Failing that, get someone who *can* follow a musical score to sit by you and mark the place.) And what does one do if, for example, a fortissimo passage is followed immediately by a pianissimo one? In this case, one or other has got to be

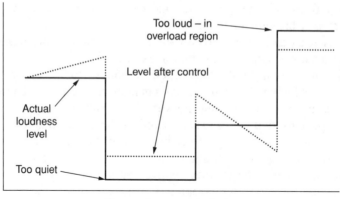

Time scale – several minutes

Figure 9.1   An idealized picture of manual control

wrong for a while and it is clearly better that the loud parts are undistorted even if the immediately following quiet music is initially too quiet. The fader can then be brought up slowly.

A point to remember here is that musicians tend to play or sing louder at the actual performance than at a rehearsal.

If a public performance is being recorded then the sound operator has to take what comes and use his or her own skills in the way just indicated. If, though, the recording is being made at the request of the orchestra or drama group and any audience is secondary in importance, it may be possible to get the leader/conductor to co-operate by letting the musicians provide some of the loudness control.

Drama is generally somewhat easier as voices do not as a rule have the same range of dynamics as an orchestra. There still remains, however, the problem of sound effects! Sounds of a gunshot are fortissimo! A clock ticking is pianissimo! It may still, nevertheless, be possible to apply the same techniques as for music. Remember, though, it is the *subjective* effect which matters. Absolute fidelity is, as far as dynamics go, less important. One could say that if, for the listener, it seems to sound right, then it is right! For instance, a gun shot in a play mustn't, for reasons we have given, be anywhere near its real loudness. But if there is plenty of die-away of the sound, which is always assumed to be a characteristic of gun shots, and in a sound-only recording the dialogue prepares the listeners for a gun shot ('My God! he's going to fire!'), then the effect can be reasonably impressive. In a television production, of course, we can *see* the gun being fired, although there are problems here because the noise of the gun may have had to be pre-recorded but played in sync with the action. (A very good application for Foldback!)

An important point here is that with digital recording, which we shall deal with later, the dynamic range of the system is much greater than with analogue recording, so much of what we have said about control may not apply. Nevertheless, if a digital recorder is used as a master but copies are going to be made on to cassettes, then control of the type we have outlined will still have to be used at some stage.

Having mentioned sound effects (and what follows applies pretty equally to sound alone or sound with pictures), it is worth making the point that these may not need to be maintained at a constant level, even though in real life they would be. Suppose a scene opens in what is clearly an urban environment with background traffic noise. It will generally be very important that the scene starts with the traffic noise loud enough to be easily identifiable. After a short while, maybe only a quarter of a minute or even less, and taking advantage of over-riding dialogue, the traffic effects can usually be taken down to a rather lower level so that they are audible but are not in any way obtrusive. It may be appropriate to bring the effects up if the dialogue suggests that this would be appropriate, and again possibly at the end of the scene.

Electronic level control

We have just shown that manual control of programme levels is often a difficult and skilled business. Properly done, though, it can be very effective. We will now deal with electronic control, which is generally far easier but if injudiciously used can have unfortunate results!

What we need is an electronic fader controlled by the incoming audio signal. There are devices which meet this requirement known as *voltage controlled amplifiers*, or VCAs for short. With VCAs the term 'amplifying' in the sense of increasing the signal voltage is irrelevant. In fact, the signal is being *reduced* most of the time. (So why call it an amplifier? The point is that the circuitry is essentially the same as in a level-increasing device.) And instead of the output level being controlled by a knob a control voltage is fed into the circuit. The simplified diagram in Figure 9.2 illustrates this.

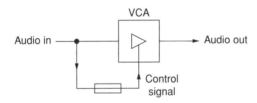

Figure 9.2 VCA controlling programme levels

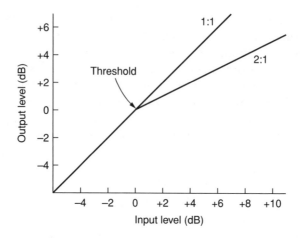

Figure 9.3 Threshold and compression

What is basically happening in the simplified diagram in Figure 9.2 is that the input level is continuously monitored by appropriate circuitry and the amplifier is adjusted to avoid over-high outputs which could cause distortion.

On professional equipment of this kind there is usually a distinction between 'compression' and 'limiting'. In either case, the VCA has no effect on the output until a certain signal level is reached. This is called the *threshold*. Above the threshold, the output is allowed to rise but not as fast as the input rises. Figure 9.3 shows this for compression.

The straight line marked '1:1' indicates that the output increases at the same rate as the input. The line labelled '2:1' represents compression and here, for every 2 dB increase in the input the output increases by only 1 dB. 2:1 is then known as the *compression ratio*. Other compression ratios will be available on the professional equipment.

Limiting is simply extreme compression, with a ratio of perhaps 20:1 or even more. Figure 9.4 shows the action of a rather sophisticated device which embodies both compression and limiting.

All this is rather heady stuff and the equipment may not be available to the non-professional user. Nevertheless, camcorders with integral microphones as well as many cassette recorders embody a form of limiter. Interestingly, the cassette recorders with this facility are often in the lower price brackets, being machines with a built-in microphone and no way of controlling the record level. Here, some method of protection against overload is essential so that fairly basic limiters are incorporated. Some professional machines also have limiters, only here they are usually capable of being switched out. They may be called something like 'ALC', meaning of course 'automatic level control'.

TERMINOLOGY
Threshold – The level at which compression or limiting starts to occur.
Compression ratio – *Above the threshold*, the number of decibels by which the input level increases for a 1 dB increase in the output level.
Limiting – An extreme degree of compression. The compression ratio will be 20:1 or more.

So far, automatic limiting seems to be an excellent idea. There is, as ever, a drawback. A further characteristic of all limiter/compressor systems is what is called *recovery time*, or *decay time*, and this needs some explanation as it is possible to ruin recordings if one is unaware of its existence. What it means is that, after the VCA has operated to reduce the level, time has to be allowed for it to return to normal (1:1). The question is, should this be instantaneous or should the recovery take a perceptible time, perhaps a few seconds? Let us take two imaginary but perfectly possible situations.

1. We are doing the sound for a videotape about the workings of a factory in which there is a drop forge. We want to record, using one microphone only, a commentary against some of the general factory noise which is not unacceptably loud except for the moments when the drop forge operates. Without any limiter the drop forge noise

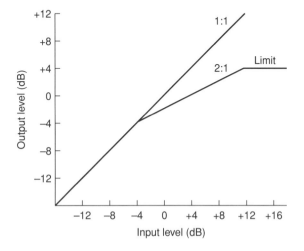

Figure 9.4 Limiting and compression

will be too high in level and consequently distorted, unless we have the recording level set so low that this does not happen, in which case the speech and general background will be too low.

A limiter will prevent the drop forge distorting. If the recovery time is very short, then levels will return to normal almost immediately after each 'drop'. If the recovery time is long, then levels will remain depressed after each drop, gradually returning to normal so that speech and background will be very low – possibly almost inaudible – for a few seconds. This is a peculiar effect and obviously to be avoided. For the drop forge factory, then, we would like a very short recovery time, perhaps a tenth of a second or less. (Even so the 'hole' in the background which the drop 'punches out' will almost certainly be audible but may not be disturbing.)

2. We wish to record a piece of music for which we have not had the chance of attending any proper rehearsal and we know that there will be some very loud passages. Again, we hope to get out of trouble using a limiter. A very short recovery time, which was satisfactory in the drop forge example, will now reduce the dynamic range of the music drastically. Loud passages will be brought down so that they may sound not very different in level from quiet sections. The result is a very boring item to listen to on playback.

If there is a long recovery time, say several seconds, the peaks will not be distorted and the slow change from limited to normal music will tend to preserve the loudness contrasts. The final result may not be ideal but neither might it be totally unacceptable.

What we are saying is that ideally there needs to be a choice of recovery times, and on professional equipment there is likely to be a switch allowing these to be selected from, say, 100 ms (one tenth of a second) to several seconds.

The small cassette machines with a built-in level control appear generally to have a long recovery time, sometimes as much as half a minute. The idea seems to be that it is probably going to be speech only that will be recorded and the limiter sets, as it were, the levels with the first few seconds of the speech.

It is quite easy to find out approximately what a particular machine's recovery time is. The probability these days is that it will be a camcorder that the reader is interested in rather than a very cheap mains-operated cassette recorder. All that is needed is to take the camcorder to a place where there is a reasonably steady low-level background noise. Traffic on a fairly busy road a few hundred metres away, or a running tap nearby, will do. A test recording lasting a minute or so should be made of normal speech which is then changed suddenly to a loud shout, perhaps from close to the camera. On playback, the background noise will be found to drop in level after the shout, coming back to normal gradually. One camcorder I tested in this way had a recovery time of about 5 or 6 seconds. This is useful information to have as it can warn against, for instance, continuing an in-vision commentary immediately after any very loud sounds. The presenter must wait a few seconds.

Advice: treat all ALC facilities with caution. They may be very helpful but don't expect perfect results from using them. On the other hand, there will be many circumstances when ALC is certainly going to be the lesser of two evils.

Recovery time – The time it takes for the signal level to return to normal after limiting or compressing.

Noise gates

These are often only found on the bigger and more expensive mixers, but are worth a mention nevertheless. We've been looking at ways of *reducing* the dynamic range, but there are also devices for doing the opposite – increasing the dynamic range. These things are called *expanders* and noise gates are examples. Figure 9.5 shows in graphical form what they do.

Suppose the input programme level (A) is reduced. At B, the output level falls by a selectable amount to C, after which it reduces at a 'normal' rate. What, you may ask, is the point of this? Well, suppose we're doing a recording in a studio where there is a little background noise – not too serious as it will be masked when there is normal speech

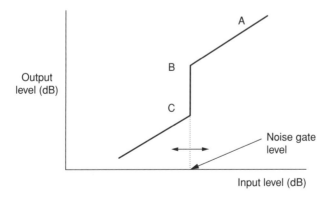

Figure 9.5 Action of a noise gate

or music, but could be audible in what should be 'silent' moments. (This can happen in television studios where there is inevitable sound from camera movements, ventilation and so on.)

If the noise gate level is set to be just above the background noise, then ordinary programme sound isn't affected, but when things should be silent the output level drops to below C and is thus very greatly reduced.

The danger is that if the noise gate is set too high then quiet programme sound also falls into the hole!

Questions

1. A limiter/compressor device is set to a compression ratio of 3:1. This means that
   a. The input rises 3 dB for each 1 dB rise in the output below the threshold
   b. The output rises 3 dB for each 1 dB rise in the input below the threshold
   c. The input rises 3 dB for each 1 dB rise in the output above the threshold
   d. The output rises 3 dB for each 1 dB rise in the input above the threshold

2. In a limiter the compression ratio is likely to be
   a. 10:1 or more   b. 10:1 or less
   c. 20:1 or more   d. 20:1 or less

10 Digital audio
Part 1

In the previous edition of this book, published in 1997 – not many years ago – I suggested that digital equipment was *at that time* likely to be rather too expensive for the amateur and semi-professional reader! Things change. Good quality digital recorders and associated devices are now very affordable.

Historical

In the UK, the first widespread application of a digital audio system came towards the end of the 1960s, when BBC research engineers developed a system for combining the television sound signal with the pictures for distribution to transmitting stations. The sound signal was converted from an analogue form into one consisting of pulses which were 'tucked into' the composite vision signal. This system, known as *sound in syncs*, is still used, not only by the BBC, but also by independent UK television companies. It is mentioned in Part 2 of this chapter.

The more modern form, known as NICAM, allows *stereo* sound to be transmitted as part of the television signal. NICAM will also be explained further in Part 2.

Then, in the early 1970s, BBC radio, faced with the problem of sending high-quality stereo to its transmitters, adopted digital audio. This not only allowed radio programmes to be sent all around the UK with almost no deterioration of quality, but financially was very economical. Use of the principles of digital audio, but with somewhat improved standards, appeared with the compact disc – the CD – around 1980, and there have been further digital systems since.

Basic principles

In analogue systems, where the electrical signal is a replica, or should be, of the original sound waves, there is one fundamental problem: any impairment caused by whatever reason is virtually impossible to correct. Such impairments could be the result of interference picked up during a radio transmission, effects of dirt and dust in a record groove, and imperfections in the recording medium (bare or thin patches in the magnetic oxide of tape, for example).

A further cause of degradation is copying. The quality of a recording on analogue tape may be very good, but a copy made from it will never be of such a high standard. Copies of the copy, a process which can be very important in the record industry, will be further degraded. With tape and suitable noise reduction processes, it may take several generations of copying before the degradation becomes very obvious, but it nevertheless does occur and is irreversible.

With the now almost defunct vinyl gramophone record, the quality deteriorated very slightly with every playing – perhaps not very obviously. The stylus gradually caused wear and deformation in the grooves and also the effects of dust were almost impossible to avoid. Hi-fi enthusiasts had electrostatic dust removers and special cleaning liquids, but the fact remained that minute dust particles still found their way into the grooves and could even be pushed into the surface by the stylus. There were various clever circuits designed to remove the effects of serious clicks and scratches, but it is true to say that they were at best only partly successful. Such a circuit could never be sure that a noise was caused by a scratch or whether it was a special percussion instrument! An amazing number of these problems are avoided by going digital.

With digital audio, the original analogue signal is converted into a code of much more robust signals. Basically, these take the form of pulses of voltage of standard height and duration. Since every part of the digital chain knows what the height and duration of a pulse should be, then if one becomes distorted, it is possible, within certain limits, to reconstruct it. There is a comparison here with Morse Code (but that seems to be on its way out!). All messages are sent in the form of dots and dashes. The person at the receiving end can usually detect and decode them, even though some may be badly distorted and that fact accounted for the widespread use of Morse for communication across great distances when interference to radio signals rendered speech difficult to understand.

What is the coding process with digital audio? Here is a slightly simplified story.

Sampling

The first stage is *sampling*. The analogue signal is split up into small samples, the 'height' (we should really say *amplitude*) of which will be measured in the next stage. A very important thing about this sampling

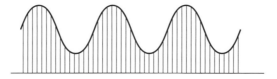

Figure 10.1   Sampling

process is that it must be carried out at a great rate. In CDs and many other modern digital systems, this rate is just over 44 000 times a second, but 48 000 is also used in some equipment. The vertical lines in Figure 10.1 represent individual samples.

Quantizing

Next, each sample has to be measured. This is known as *quantizing*.

Figure 10.2 illustrates a very basic quantizing arrangement. It shows six quantizing levels – and think of these as graduations on a ruler which measures the amplitude of each sample. The system doesn't allow us to use fractions of a division, so the measurements of the signal are going, in this case, to be rather coarse.

The more accurately this amplitude measurement is done, the better the quality of the final result. For real accuracy there needs to be a large number of graduations on this ruler. About 8000 is a usable minimum, but this is not enough for high-quality (e.g. CD quality) recordings, when about 65 000 graduations, or quantizing levels, are needed!

Sampling – The process of examining a sound signal at a very high rate. In very many digital audio systems the sampling rate is just over 44 000 times a second.

Quantizing – The process of 'measuring' the amplitude of each sample. Good quality audio needs at least 65 000 quantizing levels.

At this point, the brain starts to become bemused by the magnitude of things – measuring the music 44 000 times a second, each measurement being a number between 0 and 65 000! Fortunately, modern electronics can cope with it all fairly easily. Nevertheless, let us pause to see what we're committing the system to. A digital tape or disc must store numbers which can be anything up to 65 000 and do this at a rate of just over 44 000 of them each second!

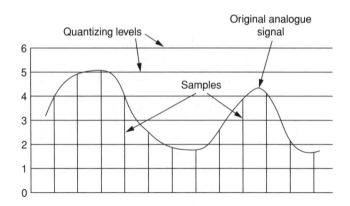

Figure 10.2 A very simple quantizing system

Binary arithmetic

Luckily, things begin to get a little simpler when we cast around for alternative kinds of arithmetic. Digital audio uses *binary arithmetic*. With our usual day-to-day arithmetic we count from 0 to 9, and then put a 1 in front to go from 10 to 19, then a 2, and so on. In binary arithmetic there are only two digits, 1 and 0, instead of ten, and the counting is from 0 to 1, and then a 1 is put in front to make 10. The next number has one added to make 11 (don't call it 'eleven' – it isn't!) and then a further 1 is put in front so that we have 100, 101 and so on. Table 10.1 compares our standard decimal arithmetic with binary. If the reader is unfamiliar with binary arithmetic, an attempt should be made to continue the table.

At first, binary arithmetic can seem very complicated. In fact, this is only because we are not used to it – at least, not most of us. In reality it

Table 10.1

Decimal	Binary	No. of bits
0	0	1
1	1	1
2	10	2
3	11	2
4	100	3
5	101	3
6	110	3
7	111	3
8	1000	4
9	1001	4
10	1010	4

With 16 bits we can go up to just over 65 000.

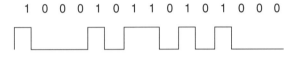

1 0 0 0 1 0 1 1 0 1 0 1 0 0 0

Figure 10.3 A binary signal

is much simpler than decimal arithmetic. The entire set of multiplication tables is:

$$1 \times 0 = 0$$
$$\text{and}$$
$$1 \times 1 = 1$$

The rules of addition and subtraction in binary arithmetic are the same as in decimal arithmetic.

However, let us see where this is getting us – we must not let a study of binary arithmetic, however brief, distract us from the main objective, which is digital audio.

After the quantizing process the resulting numbers are converted into binary, so that, for example, the number 17 832, representing a sample of about a quarter of the maximum height in a CD signal, when converted to binary becomes 100010110101000. Now, if each 1 represents a pulse of voltage we then have the signal shown in Figure 10.3.

This is the sort of code we wanted. Individual pulses can be degraded, but provided they are still just identifiable they can be reconstructed if need be.

Each 1 or 0 in digital audio is called a bit, short for *bi*nary dig*it*. Figure 10.3 illustrates a 15-bit number: there are six 1s and nine 0s. The number of bits gives an indication of the quality of the system. CDs use 16 bits. Telephone communication can be satisfactory with fewer than ten.

Regeneration of pulses

Figure 10.4 compares an analogue system with a digital one, both being subjected to interference in each section. The triangular symbols in the analogue chain represent amplifiers. The pulse generators are circuits which detect the presence of a pulse, provided it has not been too badly damaged, and create a new one in its place.

It might seem from what we have been saying that the restoration of digital pulses can go on indefinitely. This is not quite true. Eventually there will be pulses which are too far gone to be regenerated and even though most digital systems make use of *error detection and correction* – which we shall outline briefly below and explain a little more fully in Part 2 – ultimately there will be perceptible degradation. Nevertheless,

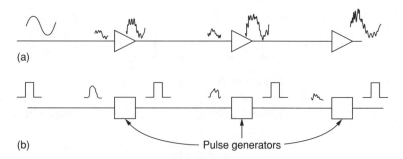

(a)

(b)   Pulse generators

Figure 10.4   Regeneration of (a) analogue signals,
(b) pulses in a digital system

digits give a far more reliable and robust way of handling any kind of signal than analogue is ever capable of. It might be worth mentioning that much of the UK's telephone system uses digital signals for distribution – for improved quality and also for economy.

Taking a 16-bit system, and with a sampling rate of just over 44 kHz (44.1 kHz in fact), it is possible to achieve a frequency range from as low as we wish up to 20 kHz. The full audio range is easily covered. And when it comes to signal-to-noise ratios, then about 100 dB can be covered. It all seems too good to be true. Not surprisingly there is a penalty and this is the number of bits per second to be handled. Some very simple arithmetic can show this.

Bit rate

With 16 bits for each sample, and about 44 000 samples per second, we have

$$16 \times 44\,000 = 704\,000 \text{ bits/second}$$

And that's for mono! Double the number for stereo and we find that we need 1 408 000 per second or 1.4 megabits/second! We shall later see that there are a lot more data to fit in, such as the error detection we have mentioned, and we can end up with over 4 million bits per second.

Now ordinary analogue tape will record up to 20 kHz, and while it's actually not quite correct to say that the frequency in hertz is precisely the same as bits per second, nevertheless the discrepancy between 20 kHz and 4 MHz is clearly enormous. Conventional tape recording simply cannot cope with digital audio. We will deal with digital tape in the next chapter, but it is enough for the moment to show that a price has to be paid for the potentially high quality that digital audio can provide.

Table 10.2

Original 10-bit sample	Number of 1s	Add parity bit?	New sample	New number of 1s
0011100101	5	Yes	0011100101**1**	6
1100110011	6	No	1100110011**0**	6
0001010110	4	No	0001010110**0**	4
1000100010	3	Yes	1000100010**1**	4

The parity bit is in **bold italics**.

Error detection

We have already made the point that with analogue systems there is no really effective way of detecting when errors of any kind have occurred.

It's somewhat different with digits. One simple but, with certain limitations, very effective method and one that is widely used is known as *parity*. What happens is this. Before recording or transmitting a sample, the number of 1s in it is counted and, if necessary, made up to be an even number. A *parity bit* is added, a 0 if there is already an even number of ones, a 1 if there was an odd number of 1s. Thus, every sample recorded or transmitted contains an even number of 1s.

Suppose an error occurs so that either a 1 is destroyed or a spurious 1 appears where there should be a 0. There will now be an odd number of 1s. So, at the replay or receiving end, when the 1s are counted and an odd number is found in a sample, the system knows that an error has occurred and steps can be taken to reduce the effect. One such step is to repeat the previous sample, and since these are only 1/44 000 second apart, the likelihood is that the deception won't be noticed. Table 10.2 may help to make the parity calculation clear. For simplicity it uses a 10-bit system.

Of course, this puts up the number of bits/second, but only by a small proportion.

Other applications of digital audio

Computer-literate readers will no doubt have already noticed that there is a big similarity between digital audio signals and the electrical pulses used in computers. In that case, couldn't computers be used with digital audio? The answer is yes. In fact, the digital editing mentioned later uses computer technology to perform its task. However, we can go further than that, and it may be sufficient here to do no more than outline some of the processing that can be carried out.

Digital delay

Just as in a computer data can be stored in memory chips, digital audio samples can be stored in chips (for as long as the power is on!). This means that with enough memory a digital signal can be fed into a delay unit and emerge from the output a short time later – from maybe a fraction of a second up to several seconds. The uses for this are many. A short list of applications is:

1. To add realism in reverberation (see under 'Artificial reverberation' below).
2. As an effect in music (not usually in classical music!) and drama.
3. To aid in 'auto double tracking' (ADT), where a single vocalist is made to appear to sing a duet. A slight delay of a few milliseconds, preferably with an element of pitch change (see below), can be very effective.
4. To improve intelligibility in PA (public address) work.
5. To enable what is called a 'profanity delay' in some radio phone-in programmes. Everything goes through a delay of perhaps about 15 seconds. If the caller uses unacceptable language the programme presenter presses a button which deletes the contents of the store, including the unwanted words. Clever electronics then builds up the delay gradually, so the fact that there has been some material removed is barely perceptible. It does, though, require a very capable and quick-thinking presenter.

Artificial reverberation

'Echo', as it is often termed rather inaccurately, was once created by loudspeakers and microphones in reverberant rooms; later, special steel plates were used and these could be set into vibration electrically to give a very passable imitation of reverberation. Both these methods were bulky and costly.

Cheaper and more compact was the spring reverberator, in which one or more springs were set into vibration by a small transducer fixed to the spring(s) with a further transducer to pick up the reverberations. Very expensive units were reasonably good but the cheap ones sounded like – well – twanged springs!

The advent of digital audio has changed all this. In a digital reverberation device, the incoming digits are stored in memory chips and then released in a very carefully controlled way, so that there is a random and steadily decreasing amplitude of the output.

This can simulate natural reverberation remarkably effectively, and there will normally be control of the reverberation time together with several other characteristics. It becomes possible, then, to produce, within quite wide limits, almost any kind of required reverberation. Modern digital reverberation units can be of comparable cost to a decent cassette machine, and take up less space. There are others, with more comprehensive facilities, which cost ten times as much, but for many purposes the cheap ones can be adequate.

Delay

One way of creating better realism with artificial reverberation is to make use of delays. This may be done with a delay unit as mentioned above, but many digital reverberation devices have variable delays built in. The reason for having a delay is this. The human ear/brain combination is very good at detecting small time intervals, such as those which occur in, say, a room when the ears receive first the direct sound from the source and then the *first reflection*, as it is called, coming from a wall, floor or ceiling. If the delay between the direct sound and the first reflection is greater than about 40 ms (=1/25 second), then one is aware of a time gap. That is the proper definition of 'echo' – when one is aware of a time gap between two sounds.

If this time interval – the initial time delay (which we've already mentioned in Chapter 3) – is less than about 40 ms, the brain does not recognize the gap as such, but is nevertheless subconsciously aware of it and uses the information to assess the size of the room. (Try blindfolding someone and taking them into a room they have not been in before. After a few moments of conversation it will almost always be found that the person can make an approximate judgement of the size of the room.) To recreate a convincing artificial reverberation, then, means using an appropriate time delay as well as the correct reverberation time. To take an example, a large hall will only be fully simulated if a longish reverberation time, perhaps 2 or 3 seconds, is used with an initial time delay of maybe 20 or 30 ms.

Pitch changing

If one imagines digital samples being fed into a set of stores but then taken out at a slightly different rate it will be seen, we hope, that the effect will be to change the pitch of the original. Figure 10.5 shows in simple diagrammatic form the principle of digital pitch changing.

In Figure 10.5, the boxes marked A, B, ..., n represent digital memory stores. Imagine the 'input electronic switch' to move round the circle,

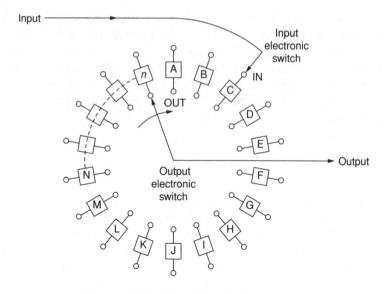

Figure 10.5   Simplified pitch changing

putting a digital sample into each store in turn. The 'output electronic switch' reads the contents of each store and then empties it so that it is ready for another sample. The samples that are read out will ultimately be converted back to analogue signals.

Suppose, though, that the output switch rotates faster than the input one. The effect will be that the pitch of the analogue output will be higher than the pitch of the input. The reader will no doubt wonder how this could be a continuous process because sooner or later one switch will catch up with, or be caught up by, the other! In fact it can't, but with the use of clever technology the impression can be given that it is continuous. The curious and puzzled reader is referred to the 'Further reading' section.

MIDI

This stands for Musical Instrument Digital Interface. This really relates to electronic music but a brief mention may not be inappropriate here.

MIDI is not a specific device, but an internationally agreed standard for connecting together two or more electronic instruments. The MIDI link carries information about the times of start and finish of a note, its pitch and other data so that, for example, an electronic organ or an ordinary PC with the right software can 'trigger' other equipment. A typical instrument such as a synthesizer may be able not only to generate MIDI data, but also to receive MIDI from another source and then pass it on to a further instrument. The book list at the end suggests suitable further reading.

Data compression

This means reducing the number of bits in an audio sample without a serious reduction in quality. It's made possible by a fairly recent development in digital audio to resort to very clever but legitimate trickery, taking advantage of the characteristics of the human ear.

To begin with there are many 'sounds' that we cannot hear: frequencies in the region of 30 Hz unless at a high level, for example. The normal ear is about 70 dB less sensitive to 30 Hz than it is at around 3 kHz and it's not particularly sensitive to the higher audio frequencies above about 3 kHz. Also, certain sounds can be 'masked' by others. It has been known for some time that a sound at a particular level can render the ear quite insensitive to nearby frequencies (usually higher) at a lower level.

By making use of circuitry that, as it were, mimics these characteristics of the human ear, it is possible to avoid recording the undetectable sounds. There are various processes. One is known as Precision Adaptive Subband Coding (PASC) or *perceptual coding*, and it makes it possible to throw away up to about 80% of the original data. Another method, used by MiniDiscs (of which more later) is called ATRAC, which stands for Adaptive Transform Acoustic Coding. MP3 also uses clever data compression.

Additional terminology

Two terms are worth a brief mention:

1. ADC: analogue-to-digital converter. This is the device (in the form of a chip) which samples the analogue signal and gives a digital output. The complementary device is the
2. DAC: digital-to-analogue converter, which translates a digital signal back to analogue.

Questions

1. The sampling rate used in compact discs is approximately how many times per second?
   a. 44 000   b. 48 000   c. 65 000   d. 704 000

2. What would the binary number 1001 represent in ordinary arithmetic?
    a. 5   b. 7   c. 9   d. 11

3. What does MIDI stand for?
    a. Musical Integrated Digital Instrument
    b. Musical Instrument Direct Insertion
    c. Musical Integrated Digital Insertion
    d. Musical Instrument Digital Interface

10 Digital audio
Part 2

Compact discs

The vital component in a CD player is the *laser*, and without this it would be impossible either to make CDs or play from them. So what is a laser? It is, of course, a light source, but the light it emits is particularly pure in that it consists of one wavelength only. Ordinary light, as is well known, is made up of a wide range of wavelengths and this causes difficulties when the light is focused with a lens. The various wavelengths are treated slightly differently and this can be seen in the colour effects when one looks through an ordinary magnifying glass. The effect is particularly noticeable near the edges of the lens. Lenses for high-quality cameras are designed to reduce this *chromatic aberration*, as it is called, to acceptable levels. The beam from a laser, though, is so pure that it can be focused with extreme accuracy with quite a simple lens.

In making a CD, the light beam from a laser is *modulated* – that is, made to fluctuate in brightness – with the digital audio signal to be recorded and then focused upon the rotating master disc. A suitable material on the disc's surface is etched by the intense beam, producing a pattern of dots corresponding to the digital signal. Accurate copies of the master can be made by taking moulds and finally extruding plastic into the moulds. The process is not unlike that used to make traditional gramophone records.

In the replay machine, a low-power laser is focused on to the track and the light reflected back is detected and converted into electrical signals. After suitable decoding we end up with a very high quality audio signal. The optics of a CD player are shown in Figure 10.6.

The diffraction grating shown in the diagram is a device consisting of very many closely spaced engraved lines which act to produce separate images of the light passing through. (With white light, a grating of this sort creates spectral colours. These are easily seen when a CD is

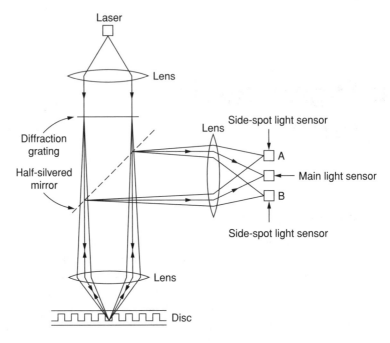

Figure 10.6   The optics of a CD player

held at right angles to the light, the tracks on the disc acting as a grating. The same thing occurs with mother-of-pearl because of the existence of fine lines on the surface.)

The reason for putting in the diffraction grating is to produce additional beams which end up at the side-spot light sensors on the right. These are used to steer the laser beam over the disc's surface. A tendency to drift is shown by an increased signal in one of the side-spot sensors and a mechanical system returns the laser to the correct alignment. How the side-spots do this is shown in Figure 10.7.

To give some idea of the fineness of a CD's tracks, these are about 0.5 µm wide (0.5 of a millionth of a metre) and the centres of the tracks are 1.6 µm apart. In simpler terms, there are around 600 tracks to a millimetre.

To add to the facts and figures about CDs, the maximum playing time is about 75 minutes. The maximum replay time is always given as 74 minutes but in fact it can be a little longer than this. CDs with a playing time of 76 minutes are not uncommon! The length of the track is roughly 5.5 km, or just under 3.5 miles! The disc is made to rotate so that the *scanning speed* of the laser is constant. This means that the data are read out at a constant rate. To achieve this, the rotational speed has to vary from about 500 r.p.m. at the inside to 200 r.p.m at the outside. We

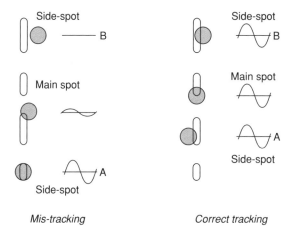

Figure 10.7 How the side-spots ensure correct tracking

put the figures in this order because the start is on the inside and the finish at the outside – opposite to a conventional vinyl record.

Error correction

Some very powerful methods of detecting and correcting errors are incorporated into CD players. These are necessary because it is impossible to guarantee perfection in the manufacture of discs on a commercial basis. One has only to hold a CD, particularly one produced at least a few years ago, in front of a bright lamp bulb to see many tiny pinpoints of light. These may be very tiny – perhaps only one hundredth of a millimetre across in some cases, but even that would represent a significant loss of data. (It may be that the manufacture of CDs has improved. I get the impression that recent CDs I have bought show fewer pinpoints of light than those bought a few years ago. It could be, of course, that I'm just lucky!)

The *writing speed*, i.e. the speed of the track under the laser beam, is about 1.2 m/s. A hole only 1/100 mm in diameter is crossed by the beam in roughly 1/100 of a millisecond. Now we have already pointed out that a stereo digital signal requires nearly 1.5 million bits/second, so even this minute pinhole would account for somewhere in the region of 15 bits.

The full story of CD (and other digital) error detection and correction is beyond this book, but suffice it to say that a good player will compensate for 'holes' of up to 2 mm in diameter.

An experiment that may be tried is to take a felt-tip pen with **washable, NOT permanent** ink and draw a radial line across the disc. When the ink has dried, play the disc and if the line is thin – less than a millimetre across – there should be no audible effects. Then make the line wider in stages until the player either mutes or repeats a track. Finally, wipe off the lines and the disc will be as good as it was before! I have demonstrated this many times in front of audiences – who are generally very impressed. An outline explanation is given on p. 150.

IMPORTANT: neither the author nor the publishers can accept any responsibility if the experiment goes wrong.

We should add that this ability to correct for missing data requires many more bits than those just needed for the audio. By the time all the other data, some for timing, some for error correction, and so on, are taken into account the final bit rate is just over *four million* per second.

Cleaning CDs

There are some very attractive looking devices on the market for cleaning CDs. Rule 1, though, is don't get the disc dirty in the first place! Keep grubby – or even clean – fingers off the surface (not forgetting that the important surface is the non-label, or underside, of the disc).

If it should be necessary to clean a disc, a damp cloth wiped care-fully *across* the tracks should do the trick. Smears along the line of the tracks could make it difficult for the system to read the data. This is why the washable ink pen in the experiment above is drawn radially – i.e. across the tracks.

Cost of CD players

There is little evidence to suggest that the sound quality from very expensive players is any better then that from much cheaper ones. There are likely to be more features on the costly ones, but whether these are important depends on the purchaser. In the early days of CDs, the error correction may have been better on the expensive machines, but this doesn't seem to be true any longer. Of greater significance is the way the extracted analogue signal is handled – the quality of the amplifiers, loudspeakers, etc.

NICAM

This stands for Near Instantaneous Companding Audio Multiplex, most of which we will try to explain! The form of NICAM used for stereo sound for television is the offspring of two separate technologies. The

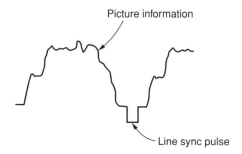

Picture information

Line sync pulse

Figure 10.8 A television line signal

first, which we referred to briefly earlier, is a method of hiding the television sound signal (originally mono) in with the picture signal, so that only one signal path is needed from studios to transmitters.

Figure 10.8 shows a typical television line signal voltage – the higher the voltage, the brighter the line as it goes from left to right across the screen.

At the end of each line, the receiver has to be told exactly when to go back to the left and start a new line – it needs a *synchronizing signal*. This takes the form of the *line sync pulse* in the diagram, and the television sound signal in digital form is fitted in where the line sync pulse is, hence the term *sound in syncs*.

The complete signal is only sent in this way as far as the transmitters. Before going out to domestic receivers, the digital signal is removed and converted into an analogue form for transmission and the sync pulses are restored, otherwise receivers would have had trouble trying to synchronize the lines.

The original form of NICAM, the other parent technology, was a method of reducing the number of bits/second. It operated almost instantaneously ('Near Instantaneous') and worked to compress the number of bits, later expanding the number when it was appropriate to do so. This process of compression with subsequent expansion is known as *companding*. The A in NICAM stands for audio – well it would, wouldn't it? – and multiplexing means combining signals together. Hence NICAM.

Stereo NICAM for television is a very ingenious method of taking stereo sound signals in digital form, applying a companding process to have a manageable number of bits, and fitting the result into the video waveform as a new form of sound in syncs. This happens as far as the transmitters. There, the stereo signal is fitted into a section of the transmitted television signal and it reaches the NICAM-equipped receivers in this form. For the benefit of non-NICAM receivers, an analogue mono version is produced at the transmitters and transmitted in the normal way.

Besides being stereo, the digital quality is much better than ordinary television sound, although not quite as good as CD quality, despite what has been claimed in some advertisements.

A little more about error correction

The reader who tried, or was tempted to try, the experiment with a water-based felt-tip pen and make marks across a CD may have wondered how the gaps were made inaudible.

With almost any digital recording system the blocks of data – that is, the 1s and 0s making up a sample – are scattered according to a universally accepted code. This is shown in Figure 10.9. The top line represents the original data in their correct order. The middle line shows the effect of scattering, while the bottom line is the data put back in their correct order.

Now suppose there has been a major loss of data – perhaps caused by a pinhole-sized gap on the disc, or some idiot drawing radial lines with a pen! This is shown by the dotted rectangle, and data blocks D, H and M are lost. However, after re-assembly, these lost blocks are well dispersed and provided there isn't too much lost, various parity checks can go a long way towards 'filling in' the missing data.

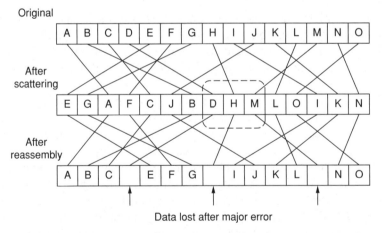

Figure 10.9   Use of a scatter code

11 Recording
Part 1 – Recording devices and systems

There are two broad categories of audio recorder:

1. *Analogue*, in which the degree of magnetization of a suitable tape varies with the amplitude of the audio signal.
2. *Digital*, where the original signal is converted into a particular code and it is the latter which is recorded, possibly magnetically.

The professional world is increasingly using digital recorders, but it is likely that for some time to come analogue machines will be useful tools for many non-professionals, so we will deal with the main aspects of these devices.

Analogue recording

Analogue machines themselves come into two categories. There is the familiar cassette type and there is what is termed the *open reel* or *reel-to-reel* type. Because of the width of the tape in the latter, it is often referred to as *quarter-inch* tape – even in a metric world! In this type of machine the tape itself is wound on spools which can vary in diameter from about 27 cm down to half that. The spools, one with the tape on it, the *feed* spool, and the other, the *take-up* spool, are locked on to suitable hubs and the tape is manually threaded through the necessary mechanism. The tape speed on such machines is usually switchable, the standards being 76 cm/s (30 i.p.s. = 30 inches per second), 38 cm/s (popularly referred to as 15 i.p.s.), 19 cm/s (7.5 i.p.s.) and 9.5 cm/s (3.75 i.p.s.). Although reel-to-reel tape machines are capable of high-quality recording, and editing of the tape is easy, they are being displaced by digital machines.

Because of their still popular usage, we shall concentrate here on cassette recorders, as the likelihood is that they will be with us for a considerable time yet. First, though, we will look briefly at the fundamental processes in magnetic recording.

The fundamentals of magnetic recording

It is well known that some substances can be magnetized fairly easily. One such substance is a particular type of iron oxide – in fact a form of rust! This is very finely ground and then made to adhere to a thin plastic strip, thus forming the tape. In any tape recorder the tape is made to move past a series of *heads*, these being basically electromagnets with coils of fine wire. One such head, the *record head*, is fed with the electrical audio signal to be recorded. The tape coating – the oxide – then becomes magnetized in such a way that the strength of the magnetization is proportional to the voltage of the original audio signal.

A second head is placed close to the record head and is known as the *replay head*. The tape with its now variably magnetized surface passes this head and small electrical voltages are induced in its coil. When these are suitably amplified we have, all being well, a signal which is a replica of the original one fed into the record head. In Figure 11.1, the letters N and S represent the north and south poles of the tiny magnets formed on the tape during recording. The left diagram illustrates a low frequency signal, where the rate of alternation is low; a higher frequency signal is shown on the right.

So that tape can be reused it is necessary to have some means of 'wiping' it – demagnetizing it – when necessary. This is achieved by a third head called the *erase head*. This is inoperative during replay, but when the machine is switched to record, the erase head, which the tape meets first of all, is fed with an alternating current of large amplitude and high frequency – of the order of 100 kHz – and this has the effect of demagnetizing the tape before it reaches the record head.

From what we have said, there seems to be the need for three heads: erase, record and replay. However, many cassette machines manage with only two – the record and replay heads being combined. They are, after all, very similar in construction. In theory, it is possible to get better results if the two heads are slightly different in their design, and this is

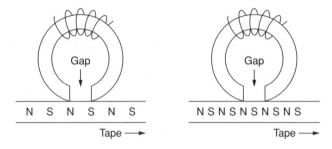

Figure 11.1   Tape passing a replay head

the case in professional 'quarter-inch' machines. There is, though, another reason, apart from quality, for having three heads and that is that the replay head can be used to monitor what has just been recorded. The more expensive cassette machines often have three heads. In such machines there will be a switch or button which in a typical case selects 'source' or 'tape'. In replay this control may be inoperative, but when set to record and switched to 'source' the signal which is being fed to the record head is also sent to meters, loudspeaker, etc.

In the 'tape' position it is the replay head which is connected to these outputs. By switching between the two it is possible to detect a delay, generally quite a small fraction of a second, this being due to the time it takes the tape to travel from record to replay head. Consequently, when recording, the condition of the recorded signal can be continuously checked for faults in the tape (very rare with good quality tapes these days), for accidental overloads causing distortion (then probably too late to do much about, but further overloads can be avoided by reducing the recording level) and for satisfactory recording in general. For example, dirty heads can result in loss of the high frequencies, a point we shall return to.

It may be worth noting that the magnetic material used on tapes has to have rather special qualities. To begin with, this material must be capable of being magnetized to a high degree. It must also be able to hold its magnetization permanently yet, at the same time, deliberate demagnetization (erasing) must not be difficult.

Cassette quality

When all things are considered the quality of recording achievable with cassettes is remarkably good. To begin with, the slow speed (4.75 cm/s – just under two inches a second) means that at high frequencies the individual 'magnets' are very short and extremely close together, but nevertheless frequencies approaching 20 kHz are recordable.

The main defect with cassettes is poor signal-to-noise ratio. The individual tracks are extremely narrow, there being four (one each for stereo left and stereo right, times two because the cassette can be turned over) in a width of just over 3 mm. And since there has to be a space between tracks then each one is less than three-quarters of a millimetre wide! Now the magnetic particles, although ground very finely, are not infinitely small, so that each one makes its own tiny and individual contribution to the recording. If the tape were only one particle wide, the result to the ear would be a series of clicks whose overall effect might perhaps have some semblance to the desired recording. The wider the track, the less noticeable these clicks become until they merge to

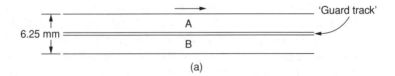

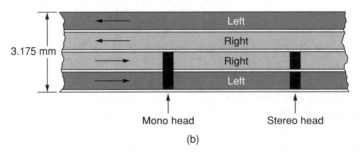

Figure 11.2 (a) Quarter-inch tape   (b) Cassette tracks

Note that if the diagrams were to scale the cassette tape would be seen to be half the width of the quarter-inch tape.

create a faint hiss, known as 'tape hiss'. In the early days of cassettes, tape hiss was a major obstacle to their serious use, but improved materials and the adoption of good noise reduction systems (see later) means that nowadays modern cassettes on good equipment can provide very respectable recordings.

The hiss, of course, sets the lower limit of the signal-to-noise ratio. The upper limit is determined by saturation of the magnetic particles. As the magnetizing force, a consequence of the voltage in the record head, increases there comes a stage where the magnetic particles are fully magnetized. Any further increases in the signal voltage are not matched by further magnetization. This condition is termed 'saturation' and on playback it appears as distortion, maybe very unpleasant distortion at that.

Noise reduction

All but the cheapest cassette machines these days, and most car radio/cassette units, have the ability to make use of 'Dolby'® circuitry. It is important to say that 'Dolby' is a registered trade name, but like many others such as 'Biro', 'Thermos' and 'Hoover' has almost become a generic term. There are three commonly used versions, 'B', 'C' and 'S'. There are others – the A system is confined to professional analogue equipment, for example. The primary aim of all the Dolby versions is to reduce the effects of tape hiss and, in the case of the cassette systems, to put it simply, they do this by increasing the high frequency content of a recording. Such recordings are often described as being 'Dolbyed' or 'Dolby-encoded', and pre-recorded ones have the Dolby 'Double D' trade mark – this can be found easily on virtually all pre-recorded cassettes.

In the replay machine a corresponding amount of high frequency reduction is used. Since tape hiss is predominantly high frequency, the result is that the second process reduces the effect of the hiss while bringing the recorded signal back to normal. In the B system there is about 10 dB of hiss reduction; with C, a somewhat more complicated version, around 20 dB reduction is achievable – even more in the S type.

To put this into a practical context, recordings taken 'off-air', i.e. recording a reasonably strong f.m. radio signal, with good quality tape and Dolby C, are barely detectably lower in quality than the incoming radio signal.

Head and tape cleanliness

A glance at the diagrams in Figure 11.1 will help to explain that for good recording and reproduction of the higher frequencies it is essential that the tape is as close to the heads as possible. It should be clear that the closeness of the poles in the high frequency case means that their effects will, as it were, cancel each other out a short distance away.

Normally, the tape is held tightly against the heads by a tension provided by the tape drive system. However, with time and use, it is easy for a deposit of the oxide material from the tape to build up on the heads, and this has the effect of keeping the tape at a distance from the heads. A loss of the high frequencies may then be apparent. The answer is to keep the heads clean. Many proprietary devices exist for cleaning the heads. A simple and very effective way, though, is to rub the heads *gently* with a cotton bud which has been dipped in isopropyl alcohol – generally available at pharmacists. (Some seem reluctant to sell this, but it can be bought as a proprietary head-cleaning material from hi-fi shops.) It is instructive to look at the cotton bud afterwards. The word 'gently', above, has been put into italics to emphasize the nature of the treatment. Rough handling of the heads can cause misalignment, which may be as bad as, or worse than, dirt!

This is assuming that it is easy to get at the heads. The better machines are generally designed so that there is access to the heads, but if this is not the case then it may be necessary to use a head-cleaning device which typically has the dimensions of a cassette and is slotted into the machine which is then set to play. After half a minute or so, the cleaning process will be finished. (Such devices are essential for car cassette players.)

It is not only the heads which must be clean, but also all the parts of the drive system and the actual tape. Normally, tapes do not get dirty themselves but careless handling can leave a thin film of grease from the skin on the exposed surface.

While dealing with the cleanliness of the heads it may be as well to include a few words about *demagnetizing*. If the metal of the heads acquires some permanent magnetism there is likely to be a change in their electric characteristics, and this can show as an increase in the hiss level. Worse still, it is possible for a tape to be permanently affected after being used on a machine with magnetized heads. Demagnetizers are readily available in good hi-fi shops and their use is easy. The instruction of both the makers of the cassette machine and the demagnetizer should be followed.

Digital recording

Having said above that digital audio cannot be handled by conventional recording machines, we will look briefly at the problem of how the enormous bit rate (number of bits per second) can be put on to tape. Let us, for the sake of simplicity, take the bit rate to be the same as the maximum frequency in Hz. As we have said, this is not quite correct, but the inaccuracy is not too serious. Let us say, then, that a conventional tape machine can handle frequencies up to 20 kHz and with digital audio we need to record frequencies up to 1.4 MHz – at least. We could do this if the tape were speeded up in the same proportion – 20 000:1 400 000, or around 70 times.

Quite apart from anything else, a reel of tape which would run for 30 minutes with conventional recording would now run for less than 30 seconds. That is faster than the spooling motors would go!

There are two tape-based approaches:

1.   The important thing, when one thinks about it, is not the actual tape speed but the *relative* speed between tape and head. Instead of moving the tape rapidly, why not move the head rapidly? This principle is used in many digital tape machines and it is also, incidentally, used in video recorders, both of the professional and domestic type, where there is a similar problem of trying to record frequencies of the order of megahertz. In such machines the head(s) are mounted in a drum which rotates rapidly. The axis of the drum is tilted slightly so that when the tape passes in front of it the recorded tracks are at a slant along the tape (see Figure 11.3).

      A compact and reasonably priced digital audio recorder with this system is known as either R-DAT (rotary head digital audio tape), or more usually simply as DAT. The original intention of the companies

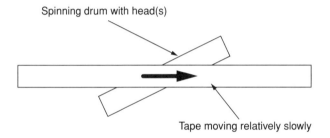

Spinning drum with head(s)

Tape moving relatively slowly

Figure 11.3 The principle of 'slant-head' recording (in practice the tape is partially wrapped round the head drum)

that developed DAT seems to have been that it would replace CDs in the domestic market, but this has not happened. DAT is, however, used extensively in the professional world.

2. Stationary head recorders. These look very much like conventional 'quarter-inch' machines. There are several models available to the professional, but they are all very expensive. Here the tape moves relatively slowly but it is of a special type which fits snugly round the heads and there may be several tracks carrying different parts of the data. Paradoxically, perhaps, the most expensive digital recorders allow the cheapest form of editing, in that the tape can be cut and spliced just like analogue tape.

DAT

1. The cassettes used are actually smaller than the familiar compact cassette, being about the same thickness but about 3 cm less in width and 1 cm less in depth.
2. The tape speed is about one sixth of that in a conventional compact cassette (8 mm/s against 48 mm/s). However, the playing time is 2 hours, with the possible option of 3 hours at slightly reduced quality.
3. The head drum inside the machine is 30 mm in diameter (just over an inch) and rotates at 2000 r.p.m.
4. A typical portable DAT machine is two or three times the thickness of this book, but its width and depth are comparable. It will run off batteries.

MiniDisc®

As we have said, it is possible, using data compression methods, to record only a small proportion – about 20% – of the sound signals that come from a microphone, and MiniDiscs do this with ATRAC, mentioned earlier. The system uses discs which at first glance look rather like computer 'floppy' discs. The rotatable disc is inside a shell. There are actually two types of MiniDisc. There are the pre-recorded ones, and these are really small CDs and are read in an exactly similar way. They cannot be recorded on!

Recordable discs are magnetic and the audio data are put on in a very clever way. The magnetic material needs to be heated to a fairly high temperature – about 180°C – to have its magnetic state changed and this is done with the laser working at high power but only on a very small part of the track. The heated part is exposed to the magnetic effects from a record head on the other side of the disc. Replay is done with the laser working at a much lower power – otherwise it might erase a wanted recording!

Now comes another very clever trick. In replay the laser beam shines on the recorded track and some of the beam is reflected, as with a CD. However, the polarization (the angle of vibration of the light waves) is changed by the amount of magnetization. This is detected using something in principle not unlike the lens of polaroid sunglasses and a suitable sensor converts this into an electrical signal. By using data reduction, which we've already mentioned, a MiniDisc can hold about 75 minutes of stereo (nominally 74 minutes) or twice that amount in mono.

A most important question about MiniDiscs is to what extent the quality is degraded, because such a large part of the incoming audio signal is discarded. The only answer I can give is that I once took a good quality CD, with a wide variety of music on its tracks. I copied a minute or two of each track on to a MiniDisc and then played each track on the CD in as near as possible sync with the corresponding MiniDisc track. (A–B comparisons, you'll note!) I switched from one player to the other, listening on good loudspeakers, and neither I nor other listeners could tell any difference! I knew which I was listening to, because I was doing the switching, but the other listeners didn't.

It is, of course, just possible that some people listening on extremely high-quality loudspeakers might just detect a difference, but I'm prepared to say that the reduction in quality on MiniDiscs is so small as to be negligible.

Solid state recording

A final point in this chapter. Since so much has been made in the sections above about storing digital audio in memory chips, it may be wondered why these are not used for *all* recording. The answer is quite simple. There is a vast amount of 'data' in even a minute's worth of sound! Let us look at it like this: we have already seen that 1 second of stereo contains about 1.4 megabits, and this is without taking into account bits for timing, error detection and correction, and so on. Two megabits per second would be a reasonable bare minimum. For a minute's duration of high-quality sound, this means about 120 megabits. Now the unit of memory storage in computers is the byte, usually equivalent to 8 bits, so a minute of good stereo needs around 14 or 15 megabytes.

Only a few years ago this would have been a total impossibility, and with conventional chips it's still not all that easy. However, two developments have taken place. First various 'memory cards' with many megabytes of storage have appeared. (Digital cameras use 'flash cards' with many megabytes of storage.) Secondly, there have been advances in 'data compression' (see below). This has led to the appearance of MP3 recorders, which can store long periods of quite acceptable music with no moving parts.

In the professional world there have been some 'solid state' recording machines but they, and the 'blanks', are currently very expensive. It has to be said, though, that high-quality recording and reproduction with no moving parts is a very attractive idea. A large part of the cost of a high-quality tape recorder is in the rigid but preferably not too heavy deck to avoid mechanical distortion of the tape path. There is also the necessity of having very accurate speed control of the tape without any risk of stretching it when spooling and braking. If all that can be avoided…!

There is, however, what may be the start of affordable solid state recording and that is MP3.

MP3

This is a now well-known recording device. By using heavy data compression, the machines can record music for considerable durations on to a solid state storage card. The quality? Well, it's difficult to say because the system is designed for headphone listening and this makes it difficult to make A–B comparisons with anything else.

Questions

1. In a three-head tape recorder, what is the order in which the tape meets the heads?
    a. Record, erase, replay   b. Replay, record, erase
    c. Erase, replay, record   d. Erase, record, replay

2. At what speed does the tape travel in a cassette recorder?
    a. 4.75 cm/s   b. 9.5 cm/s   c. 19 cm/s   d. 38 cm/s

3. What is the likely effect on the reproduced sound signal of dirt on the heads?
    a. Reduced high frequencies   b. Reduced low frequencies
    c. Increased distortion   d. Damage to the tape

11 Recording
Part 2 – Editing

Why edit?

There are two general reasons for wanting to edit a recording:

1. To shorten or rearrange the order of items on the tape.
2. To 'clean up' a recording by removing things likes coughs and other accidental noises. Under this heading can come mistakes such as mispronunciations by a speaker who has immediately corrected them, excessive 'ums' and 'ers' and general 'fluffs'. With 'quarter-inch' tape, the mechanics of editing are fairly easy. A skilled editor can remove the smallest imperfection, helped by the fact that such tape may be moving at 38 cm/s, so that an unwanted sound lasting only one fiftieth of a second will occupy several millimetres of tape.

In the case of cassettes it can be said outright that any kind of editing is far from easy! In fact, it should be discouraged if only because a break at an edit point could cause a lot of mangled tape to get caught up in the mechanism.

I know of a Talking Newspaper for the Blind studio which gets round the problem by transferring cassette material to 'quarter-inch' and then editing that. The edited version is finally put back on to cassettes for distribution.

If such facilities do not exist then what advice is there? Let us take first the rearrangement of items. With a second cassette machine this should not be too difficult. Items from the first ('master') tape are copied on to a tape in the second machine in the required order. The process may be a little tedious but should be effective. There is, though, the risk of detectable degradation of quality in the copying.

What we may call 'fine' editing – the cleaning-up of short duration mistakes – is virtually impossible unless copying to a large format tape can be done. In other words, the options that are open to the recordist are either to do a retake or live with the mistake. The danger with retakes is that the performer(s) tend to be more nervous and this increases the risks of further and possibly worse mistakes! The question has to asked whether the 'fluff' is important and it may be worth making one or two general observations about such 'mistakes'.

1. It may be desirable to keep in some 'ums' and 'ers' if they are a characteristic of the person being recorded and provided that they don't become a distraction. For example, and this is purely hypo-thetical, the Chairman of the Governors of a school is speaking at the official opening of some new school buildings. He is well known for making slight, possibly nervous, coughs at fairly frequent inter-vals. It might be better (besides being easier!) not to try to edit all the coughs out, as this gentleman would sound unnatural without these characteristic noises. Of course, the editor has to make a prob-ably not-too-easy judgement – how many coughs can be left in before they become distracting? Also, is there a risk of making the whole thing unnatural because there may be slight inflections in the voice which occur after a cough and would be out of place if there were no cough. Trial, error and experience have an important part to play!
2. When considering whether to edit out anything, the background noise has to be considered. If this fluctuates in any way then there is a risk of a sudden and unnatural change in the background where there is an edit point.
3. Although one should always expect – and possibly hope for – listeners to be critical, there is no point in worrying about little noises that will probably not be heard. It is an almost universal truth that if the subject matter is interesting so that listeners (and this includes listeners to the sound accompanying television pictures) are engrossed in it, then minor flaws can go unnoticed. This, of course, is no excuse for sloppy work!

Practicalities

There are two aspects to editing:

1. First there is the mechanics of the job – the process of removing an unwanted sound, moving one section to somewhere else, and gener-ally improving (hopefully!) a recording.
2. Choosing the exact point, for example, to remove an unwanted sound; how much of what to move where. This is perhaps what we'd call the 'artistic' side of editing. This is the more difficult part. Some would say it's by far the more difficult part.

So we will try to deal with both, but it's not easy to show how to deal with artistically satisfactory editing in a book!

Editing tape

Full-size tape has the advantage that it is quite easy to cut and join sections. An 'editing block', which is a rather precisely machined metal bar with a slot cut in it lengthwise, is required. The slot is shaped so that it has a slight gripping effect on the tape, to hold it in place. Slots are cut in the block, usually at 45°, 60° and 90°, to act as guides for a single-sided razor blade. Precision cutting of the tape is thus possible.

Then to join two pieces of tape they are put into the block, butted up to each other and joined with a short piece of special tape.

CAUTION. Used razor blades must always be disposed of carefully. A carelessly left blade could easily cause someone to have a nasty cut, and there could be worse hazards: AIDS, for example.

Cassette tape is almost impossible to edit by cutting. Instead 'dub editing' is the easiest option. This means copying on to another machine and assembling wanted sections of the recording in whatever order one wants. Unfortunately copying of cassettes results in a degradation of quality and very fine editing is extremely difficult to do.

Digital tape editing

The trouble with digital recordings is that editing is a complicated business. It is quite impossible to cut the tape, as is easily done with conventional tape. The only method of performing edits is to transfer the digital signals on to a computer hard disk and then suitable software allows the operator to remove and reposition sections of the recording. A bonus is that the editing is 'non-destructive', meaning that the original is not altered.

Up to a few years ago, such digital editing systems were very expensive. However, with the hard disk capacities and processing speeds available on ordinary PCs, digital editing packages are now quite affordable. At least one of the books listed in 'Further reading' can provide much more information on this subject.

MiniDisc editing

A moderate amount of editing is very easily done with a decent machine. Sections of a recording can be marked off as separate tracks. Suppose a part of the recording is to be removed. The start and end of that part can be 'marked' so that the unwanted section appears as a complete track, which can be erased. A degree of moving 'tracks' can also be done.

It is doubtful whether it's possible with MiniDisc to do the very fine editing that's possible with full-size tape, where, as we've said, a good editor can take out fractions of a second of a recording by cutting. Similar accuracy is achievable with computer editing. Nevertheless, it's not difficult to edit to within half a second with a MiniDisc player's own system.

12 Public address
Part 1

It would be as well to start with a clarification of the term 'public address', or 'PA' for short. It obviously means having a system of loudspeakers placed so that the output of microphones or other sources can be fed to a relatively large number of people. It is, though, convenient to realize that there are three quite different applications:

1. The audience is remote from the microphones. A typical example is that of a sporting event where the commentator is usually in a small room some distance from the loudspeakers. Another good example is on railway stations where the aim is to inform people of train arrivals, delays and so on. In both cases intelligibility is (or should be!) the most important factor. Sound quality is a secondary consideration.
2. The audience and the microphones are in the same room or hall. The term *sound reinforcement* would be better here than 'public address' because that is usually what it is. In other words the audience, or at least much of it, can hear the performers but probably with a degree of difficulty. What is wanted is reinforcement. Perhaps unfortunately, because it isn't quite accurate, this kind of situation is usually referred to as 'public address' or 'PA', and that is the term we shall use here, simply to follow generally accepted practice.
3. The amplification of music at a rock concert and similar events. While very important in the entertainment world, this is not a topic we shall try to cover here as the amount of amplification and the scale and size of the equipment needed is not likely to be within the range of interests of the reader. However, many of the principles we outline here will still apply.

The first type of PA can be dealt with quite briefly, as installations are usually permanent and normally the only problem is of having enough power. To avoid confusion we will call the two situations by their likely environments – *outdoor PA* and *indoor PA*.

Outdoor PA

In many ways this is the easier of the two, as there will normally be little or no risk of the microphone(s) picking up significant amounts of the loudspeakers' output, this being the major problem in indoor PA.

A big problem one frequently finds in outdoor PA (and this is sometimes very noticeable on large railway stations!) is that the sound level from individual loudspeakers has to be high to make sure that there is adequate loudness everywhere. Unfortunately, this is apt to mean that the sound levels near the loudspeakers are uncomfortably loud. Also, at some positions there can be annoying multiple repetitions of sounds caused by different distances from the speakers. The solution is usually to have a large number of small loudspeakers, each one producing a correspondingly lower volume. It must be admitted, though, that this piece of advice is sometimes one that for very good reasons cannot be followed: the cost of large numbers of units may be prohibitive, or it may not be possible to find the right positions for them all to be mounted safely.

There are likely to be very long cables to the loudspeakers and there is then a risk of loss of power in the wires. A common answer is to use what is termed 100-volt working. Briefly, this means that the system copies the national grid in a miniature way, where power is distributed at high voltage and low current. Part 2 explains this more fully.

Indoor PA

The problem here, almost invariably, is that the microphone(s) and loudspeaker(s) are not very far apart so that it is easy for the microphones to pick up some of their own amplified output. This then goes round the system again and there is a grave risk of an oscillation, resulting in a (usually) high-pitched squeal. This is often called 'feedback'. A more descriptive term which we shall use here is 'howl-round'.

Howl-rounds can be most annoying to an audience, as well as extremely embarrassing to the sound operators! Part 2 explains a little more fully their causes; at this point we will simply suggest some remedies to try. They may not all work!

1. Reduce the sound levels from the loudspeakers to the lowest that can be accepted. This may not be as drastic as it appears, as it can often happen that the audience near the stage/platform receives a

reasonable level of direct sound and therefore little reinforcement is needed.

2.   Angle loudspeakers away from the microphones. This can help but if a lot of the returned sound is via a hard wall then angling may have only a small effect.

3.   Use directional microphones. Cardioids with their dead sides towards the auditorium are an obvious solution. If there have to be loud-speakers close to the stage it is possible that hypercardioids, with their two 'dead' regions, may be more effective.

4.   If EQ is available on the amplifier or mixer being used then judicious use of it can help. To be really effective, however, the EQ needs to be comprehensive. For example, the use of presence cut at or near to the howl frequency is likely to be much more use, and possibly cause less deterioration of the wanted sound than simple top or bass cut. (A parametric equalizer, *if available*, can be very useful here.) Consider, if possible, 'line-source', sometimes called 'column', loud-speakers. These are stacks of drive units, often an odd number like five or seven, in one cabinet and they have a marked directional effect, albeit with limitations. Briefly, it can be said that if we suppose such a stack to be vertical then the sound radiated from it tends to be confined to a relatively small vertical angle but a wide horizontal angle. If this type of loudspeaker is used so that, for example, it is above the microphone, then there can be quite high sound levels out in the audience area but very little at the microphone.

Some general comments about indoor PA are the following:

1.   There is a great tendency for PA levels to be too high. The performers in front of the microphones are themselves often to blame for this! Many feel that they are not being heard unless they can hear them-selves clearly from the PA loudspeakers. Admittedly, it is possible for a speaker to get great reassurance from this, but restraint should be tried.

2.   Personal microphones clipped to the clothing usually seem to give reasonable immunity to howl-rounds. As they are generally omni-directional, the reason for this is slightly puzzling. Absorption and/or shielding by the clothing and body of the wearer may help to explain the effect. Additionally, the fact that the microphone is a fairly constant distance from the mouth makes the sound levels more predictable.

12 Public address
Part 2

Howl-rounds

Figure 12.1 shows in diagrammatic form the fundamentals of a PA situation. The amplifier feeding the loudspeakers raises the signal level, while there are losses in the microphone, the loudspeaker and the room itself.

If the gain (amplification) is greater than the total losses then there will be a howl-round. However, if the amplifier is set so that this is not in general the case there can still be trouble if, at any frequency, there is an extra gain caused by the addition of the frequency responses of the main components.

Figure 12.2 illustrates what might be typical frequency responses of the main components in a PA 'loop' – the chain which is capable of causing a howl-round. Those components are:

1. The acoustics of the room itself.
2. A loudspeaker.
3. The microphone.

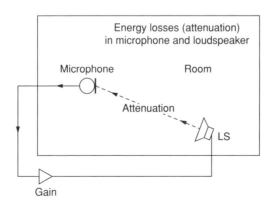

Figure 12.1 A basic PA situation

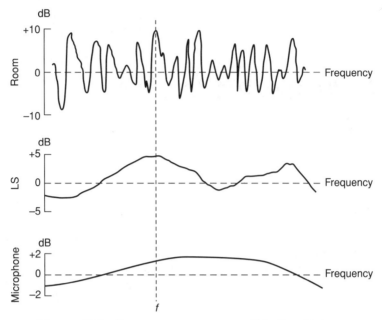

Figure 12.2 Frequency responses within the PA loop

In the diagram it can be seen that at the frequency *f*, marked with a dotted line, all three components have a response which is higher than the average. In other words, at this frequency there is in effect additional amplification, by perhaps as much as 15 dB. It is at this frequency that the additional 'gain' may result in a howl.

Moving the microphone can create different conditions and the howl may go away, or get worse, or even occur at another frequency! Changing the microphone position is nevertheless worth trying. If the microphone is to remain static, then an equalizer producing a dip in the response at *f* will almost certainly give a worthwhile improvement.

100-volt systems

This is something of a misnomer as it suggests that the voltage involved is 100; in fact it is only *notionally* that value. The idea is simply that transformers are used at the output of the amplifier which feeds the loudspeakers to step up the voltage to, conceptually, 100 V. What the voltage actually is depends on the signal at any instant. When the speech, or whatever, is quiet then the line voltage will be much less than 100 V, obviously.

To take a very much simplified example, suppose the amplifier is delivering 10 A at 10 V. A long run of cable to loudspeakers might have a resistance of 1 Ω. But the power produced as heat in a resistance is given by:

$$P = I^2R$$

where I is the current in amps and R is the resistance in ohms.

Now, with 10 A and 1 Ω, the power lost in the cable will be

$$(10)^2 \times 1 = 100 \text{ W!}$$

If true, this suggests that all the amplifier's power is lost as heat. Clearly this cannot be the case, but this is a simplified example and what will be true is that a substantial amount of power will be lost as heat in the cables.

The reader will probably be well aware that a transformer which steps up the voltage in a particular ratio steps the current in the same proportion. So if the amplifier's output goes into a transformer which steps the voltage up ten times (in this example to a nominal 100 V), the current will go down by a factor of 10 – from 10 A to 1 A.

The power loss in the cable will now be

$$1^2 \times 1 = 1 \text{ W}$$

– a power saving of 99 W! (Not really this much if the calculation were done correctly, but there will be a substantial saving.)

Of course, there must be transformers in each loudspeaker which restore the voltage and current to values more appropriate for the unit. It is common for these transformers to have variable tapping (the step-up/step-down ratios can be changed) and this means that the outputs of individual speakers can be preset. Even a complex system can then be optimally balanced even though there may be only one power amplifier.

Questions

1. The principal radiation of sound from a vertical column loudspeaker is
   a. Wide horizontally, narrow vertically
   b. Narrow horizontally, wide vertically
   c. Narrow both horizontally and vertically
   d. Wide both horizontally and vertically

2. A person is going to speak to a fairly large audience in a hall. PA is needed. Which microphone from the list below is most likely to be suitable?
   a. Omnidirectional   b. Figure of eight
   c. Cardioid   d. Hypercardioid

13 Music and sound effects

A good choice of either of these can greatly enhance drama or documentary recordings, whether for sound only or to accompany pictures. Inept choices can ruin the effect. Let us take music first.

Music

First of all, why does one want any music? Having asked the question the really honest answer might be that one doesn't, which solves the problem. However, there are many reasons why one may want music. Some of these are:

1. To establish a particular mood or atmosphere.
2. To fill in 'blanks' where there is no commentary – but at the same time it may be advisable to avoid the habit (fashionable amongst many television producers) of avoiding any silence at all costs!
3. If the programme is to be one of a series then it may be appropriate to have some introductory music as a kind of signature tune. This could be particularly true in an educational system, where all programmes dealing with a specific range of topics could all be identified with the same music.

Then, having perhaps decided that music of some sort is needed, the next and most critical decision is the music itself. The following can be no more than a guide to this choice:

1. Avoid using one's own favourite, simply because you like it!
2. The use of current numbers from the charts will quickly date the programme.
3. The music should be appropriate. Decisions here can be very difficult, but to give just one example, a delicate violin sonata would not match a documentary programme about heavy engineering; heavy

symphonic music might be better. And if the programme is about, say, a repetitive mechanical process with its own rhythm, then any accompanying music obviously ought not to have a rhythm that doesn't match that of the machinery. Introductory music could, of course, have its own rhythm which is not necessarily related to that of the machinery.

4. Be careful about accidental puns or other inappropriate linkages in the title. Haydn's 'London' symphony would not do for a programme about Manchester. And, if one may use a facetious example to make a point, a zoological documentary about insects would sooner or later cause unwanted amusement if music from the Beatles were used!

5. If using pre-recorded music be careful about copyrights.

One other very important factor affecting the choice of music, and one that it is very easy to overlook, is the length of the piece before a suitable fade-out comes. Suppose there are 30 seconds of opening introduction, either of speech or pictures. Then the music item should be such that a neat fade-out can be made after about the 30 seconds. How one defines a good fade-out point is very difficult. It can be demonstrated, but even then not always very easily. A fade should be clear, and that means that if the music is getting quieter, or descending in pitch, any superimposed fade is going to lose its impact. It may be well worthwhile spending some time listening carefully to the use of introductory music on the radio or television when usually (but not invariably!) the fades are handled very well.

It may be that the music needs to be continued under speech, 'under' meaning 'quieter than the speech but still audible'. This is an additional restraint. The balance between speech and music needs to be carefully judged, and the music should not be of the kind where the listeners/audience are going to be straining their ears to hear the tune and not concentrating on the voice.

As an example of a thoroughly inept choice of music, I once attended a very good amateur performance of a play set in Classical China. ('Lady Precious Stream' – not performed so often these days.) The intro music was the opening of Beethoven's Fifth Symphony!

Sound effects

Probably the first question is, again, are sound effects needed? The answer may well be yes, but if it happens to be no then one should resist the temptation to put in effects for the sake of it. Good, meaning appropriate, sound effects can greatly enhance a programme. Poor and unnecessary ones can be a most undesirable distraction. The makers of the programme have to discipline themselves – avoid any trace of self-indulgence when deciding about sound effects – or all other aspects of the production, in fact.

Sound effects can fall into one of three categories:

1. Naturally occurring background sounds.
2. Specially recorded effects to be added either at the time of recording or later.
3. Pre-recorded effects from, say, commercially available CDs.

Care needs to be taken with all three. We will consider each in turn.

Naturally occurring background sounds

In one sense these are the most difficult to deal with. The implication is that the microphones in use pick such sounds up whether one wants them or not. At first sight one may feel that these noises are bound to be correct because they were part of the sound scene at the time. It is perfectly possible, however, to find on listening to a playback that they are not appropriate.

The following is a hypothetical, but nevertheless perfectly plausible, situation.

We are recording a video programme in a small country town. There is a presenter speaking to camera. A nearby church clock chimes the hour. Now:

1. We could assume that the viewers will accept that there is a church nearby even if it is not in the picture. But what about viewers who do not know the locality?
2. We could make sure that the church is in the picture. This gets round the first objection but then the question is: what relevance has the church to the programme? It may have nothing to do with the programme content, in which case should we not avoid the chimes entirely?

(There is a well-known and very powerful law which lays down that if we *wanted* the chimes then the day of filming would be the day when they were out of order!)

On the other hand, there are many natural sounds which are going to be essential to help establish the environment. Imagine a presenter in a farmyard without background sounds of cows, poultry and so on. But, again, be careful. If there are pictures then the cow and poultry noises *must* be reasonably relevant to what is seen. If it's a dairy farm, sounds of pigs and poultry would probably not be appropriate.

Specially recorded effects

Inevitably these are going to be mixed in with the actuality sound. Usually, someone goes off with a portable recorder to an appropriate location some time beforehand. There is a warning here: it can happen that a recording of the actual effect may not sound a bit like the thing it is supposed to represent. Falling rain can be a good example. Recordings of actual rain frequently sound like a hiss. What is often needed is something which includes the sound of individual drops mixed in with a more general background and this can be obtained in the right circumstances. It was not for nothing that the standard BBC recordings of rain were, for many years, achieved by rolling rice around on the top of a drum! This could often sound much more realistic than the real thing!

The only advice here is to consider very carefully the nature of the sound effect wanted and be prepared to do many takes before settling on the version one wants.

Commercial pre-recorded effects

These have become much more readily available in recent years, although the cost may be too high for some budgets. Generally they are very good but, as ever, there are warnings. The main one concerns suitability – in two possible senses.

If it is to be a sound-only recording the effect must be clear and unambiguous. This is obvious, *but it may be worth stating nevertheless.*

With pictures there is a quite different hazard. Since the viewer can see the environment, then it may be that the clarity and unambiguousness is allowed a little latitude – unless the apparent source of the sound is in the picture. Let us take an example, again hypothetical but plausible. We are filming a steam train and on the day there is too much wind, and in any case the train is some distance away, so the microphone's pick-up of the train is quite inadequate. We decide that the way out of the problem is to add a commercial recording of a steam train at a later stage. So, after trying with some success to match the level and perspective of the mixed-in sound to the apparent distance of the train in the pictures, we labour for hours trying to synchronize the puffing noise with the pictures of the smoke coming out of the chimney. Our delight in our achievement is crushed when a train enthusiast points out gleefully that what we saw was a Southern Railway 0-6-0, whereas what we heard was a Great Western 'Castle' class locomotive!

The same sort of thing can happen with aeroplanes and motor cars!

To summarize sound effects in two basic rules:

1. Be very careful about the appropriateness and apparent naturalness of the effect.
2. Do not allow the level of the effect to be distracting. It is usually possible to establish the nature of the effect and then take the level down below any speech so that there is no more than a suggestion of the effect. At the end of the speech, the level of the effects can be brought up.

14 Safety

Safety is sometimes seen as a boring subject that one pays lip service to if someone else is watching! Health and Safety at Work legislation in the UK has done much to increase safety awareness in industry, maybe not before time, and this book would be failing badly if it did not outline aspects of good safety practice. We shall mention four areas – electrical safety, noise hazards, general 'mechanical' safety and fire.

Electrical safety

We are so used to electricity that sometimes familiarity breeds contempt and it is possible for there to be fatal consequences. All mains-powered equipment should be regarded as potentially hazardous, yet the rules for avoiding danger are really very simple.

First of all, the mains plug must be wired correctly – the brown wire to the 'L' (live) terminal, the blue to 'N' (neutral) and the striped green/yellow to 'E' (earth). The latter is vital for safe working. The practice of removing the earth connection to reduce the pick-up of mains hum is a thoroughly bad one.

A most basic rule is never to open up the casing of equipment without first ensuring that it has been disconnected from the supply. And one should never rely on anyone else to do the disconnection without checking. It is all too easy for the other person to mis-hear an instruction to remove the plug from the socket. Besides this obvious rule there are two important safety devices which, properly used and understood, can give essential protection.

Fuses

These protect the equipment and not the person – at least not directly. A correctly rated fuse will prevent an item of equipment from carrying too high a current in the event of a fault causing a short circuit.

To take a domestic item as an example, a typical small hair dryer is rated at about 500 W. This means that it nominally takes about 2 A. The majority of mains fuses for fitting into plugs are rated at 1, 3 and 13 A, meaning that the thin piece of wire inside them will melt and break the circuit if the rated current is exceeded. The hair dryer will clearly need a 3 A fuse. Suppose now that the device develops a fault. As soon as the current exceeds 3 A the fuse will 'blow' and no current will flow. If this did not happen, the excessive current could cause the dryer and its cable to overheat with a serious risk of fire, or at the very least, considerable damage to property. It is therefore most important that all electrical items are correctly fused. (Mains plugs are frequently supplied with a 13 A fuse fitted. This is not really a very good idea. It might be better not to provide a fuse but supply instead a little card stating the correct fuse ratings for a range of household objects. It would surely cost no more – possibly less!)

Fuse ratings

Power (watts), current and voltage are related by

$$Power = Current \times Voltage$$

Assuming that the supply voltage is 230 V – normal in the UK (220 V is more common in Europe) – then the current taken by a piece of equipment is

$$power\ in\ watts/230$$

It is good enough in practice, and it makes the arithmetic much easier, to take the supply voltage as 250 V. Hence a 500 W item takes about 2 A, and so on. The power rating of most audio equipment is low – below 200 W – and therefore it might seem that a 1 A fuse would be correct. Unfortunately, many pieces of apparatus have an initial surge current, when first switched on, which means that, in this case, a 3 A fuse would be needed.

Increasingly, and rightly, manufacturers state in their literature what fuse to use. The important thing is not to go to the fuse rating higher than the recommended one.

Note, though, that a fuse does not directly protect the person. If one were silly enough to touch the bare wire of a mains cable, the likelihood is that death would occur long before any fuses blew, the simple reason being that currents far less than even 1 A can be lethal.

There are two main aspects of an electric current which can be dangerous. One is the actual current (in amps) and the other is the time for which that current flows in the body. Of course, much depends on the route through the body. A current flowing near the heart, for instance, is likely to be more dangerous than one through the hand only.

Circuit breakers

These fall into two categories: Residual Current Devices (RCDs) and Earth Leakage Current Devices (ELCDs).

The most dangerous fault which can develop in equipment is that in which exposed metal becomes 'live'. If the fusing is correct, this should result in an immediate blowing of the fuse. However, this is by no means certain. Touching the live exposed metal can be especially dangerous if one is also in contact with another, properly earthed item.

I was once involved as an expert witness in a tragic instance when an electric guitar had become live through faulty plug wiring and the musician holding it had touched an earthed microphone stand.

Out of doors the ground itself will be, very literally, at earth, and the consequences of touching live metal may easily be lethal. Almost certain protection is provided by devices of the 'Power Breaker' type. These are plugged into a mains socket and the cable to the equipment is in turn plugged into the device. Units of this sort detect any kind of leakage of current to earth and they interrupt the supply if the current exceeds (usually) 30 mA – about 1/30 A. Furthermore, the British Standard requires the device to operate within one thirtieth of a second. This is considered to give adequate protection against shock resulting from a live/earth current. It does not guard against a live/neutral flow – in other words, getting hold of both the 'live' wires. Nothing will protect against that!

The use of RCDs, or their equivalents, should be seen as absolutely essential whenever mains equipment is used out of doors, for example with PA equipment at a sporting event. Their cost is not great and when not being used for audio purposes they can find a place in the home, giving protection when using hedge-trimmers and electric lawn mowers!

Isolating transformers

It can be argued that these give the best protection as they totally isolate the apparatus from the mains. They are, however, bulky and expensive, and only *one* piece of equipment must be connected to each transformer. (The transformers used in the construction and building industries for use with heavy duty electric drills and the like are usually not true isolating transformers.)

Checking the equipment

So easily put off till another day, this is very important. With equipment that gets moved around, it's more than important – it's vital. A simple check list includes looking for frayed leads, cracks or other defects in the insulation, cracked plug covers, wires firmly held in their terminals, and so on. Regular inspection should be seen as being at least as important as achieving a good recording!

A warning

Many large buildings, theatres and concert halls, and also industrial premises, have more than one phase of the mains present. Without trying to explain this, we will simply say that this can result in much higher voltages than 230 V between items of equipment or machinery connected to different phases. If in doubt, consult the resident electrician or safety officer.

Electric shock

Perhaps this is being a bit morbid, but what do you do if someone gets an electric shock and is lying on the ground unconscious – possibly dead? We all hope this never happens, and I think it's fair to say that incidents of this sort are pretty rare in recording and broadcast studios – BUT!

There are recommended procedures for giving resuscitation, and this book isn't an appropriate place for dealing with them. However, easily obtained wall charts outlining the methods exist and they should be displayed in all technical areas. It's very important that everyone who works in that area knows what to do if the awful situation occurs.

Noise and hearing

This can be a contentious topic. All we will do here is to summarize current UK legislation. Basically, there is a fairly severe risk of permanent hearing damage if the equivalent sound level, denoted by L_{eq}, exceeds 90 dB(A) for 8 hours in each working day.

Before going further we should try to say what is meant by L_{eq}. It is a kind of average of the sound energy which enters the ear. The full definition is rather more complicated, but a measurement of L_{eq} takes into account not only the sound level, but also its duration. Simple sound level meters are unlikely to be able to measure it; what are called *integrating meters* are needed and these are often of the more expensive

Table 14.1

L_{eq} dB(A)	Duration
90	8 hours
93	4 hours
96	2 hours
99	1 hour
102	30 min
105	15 min
108	8 min
111	4 min

kind. However, assuming that there is some way of determining L_{eq}, the relationship which is generally accepted between it and the duration of exposure if the risks of hearing damage are to be minimized is shown in Table 14.1.

It will be noticed that the permitted exposure time halves for each 3 dB increase in L_{eq} (a 3 dB increase is equivalent to a doubling of the power).

It may be reassuring to say that in normal circumstances there is little likelihood of hearing damage for the person who is not working close to high power loudspeakers for more than a very few hours a week. As an example, a typical professional monitor loudspeaker may be able to produce around 120 dB(A) at a distance of 1 m in front of it. This, note, is the peak power from it. In an ordinary session, the levels will probably be well below that for much of the time, and there will also be rehearsal breaks, and so on. Also, there may be no need to be as close as 1 m from the loudspeaker. At a distance of 3 m, the level will probably be 8–10 dB lower than at 1 m.

Consequently, the L_{eq} over an 8-hour day could easily be below 90 dB(A).

This is not to encourage complacency! Far from it. One should always be aware of the risks which can arise from high sound levels. At the same time, we are trying to avoid being alarmist. What I find worrying is the number of people who drive around with their car windows closed and the radio or cassette player operating at a level which makes it audible some distance away in a busy street, and those wearing 'walkmans' which can be heard several seats away on a tube train!

'Mechanical' safety

This is not a very good term, but one that may suffice to cover non-electrical, non-acoustic aspects. A great deal of what follows is obvious common sense, but quite often the obvious needs to be pointed out before it becomes obvious.

The following list is far from comprehensive, but it may be enough to help the reader to think of other aspects of safety.

1. Scaffolding. This is sometimes used to support sound or lighting control areas in theatrical productions. Properly assembled scaffolding is very safe, but if in any doubt an expert should be consulted. Luckily there are, in any community, likely to be builders with experience and knowledge. All platforms must have safety boards around their bases to prevent feet from accidentally slipping over the edge and also to stop things like screwdrivers rolling down into the audience area. Ladders must be firmly lashed and at not too steep an angle; 75° is about right, equivalent to a slope of 1 in 4.
2. Any equipment suspended over an audience must be well secured. Also, it may be against local licensing laws to have microphones hanging over an audience. It is worth checking.
3. We have already mentioned the importance of keeping cables in positions where no one can trip over them. As a reminder:
   (a) Take cables over doorways if possible.
   (b) Where cables have to be on the floor, then either cover them with mats or carpets or tape them to the floor so that loops, which could catch shoes, cannot form.
   (c) With things like table microphones, the mic cables should be tied firmly round the leg of the table as close to the floor as possible. This prevents the cable rising too high if it is accidentally pulled. It can also avoid strain on the cable entry into the microphone and save the microphone from being pulled off the table.

Heavy weights?

Dealing with heavy things can come under 'Mechanical safety'.

Question: What do you do if there's a heavy item to be lifted?
(a) Show that you're tough and swing the thing up in the air, especially if there's an admiring audience.
(b) Use some sense.

(b) is correct!

There are two possibilities: either you can lift the thing yourself, or you can't.

If you can, then the important thing is to do so without injuring yourself. And what, in these circumstances, is the most vulnerable part of the human body? It's the back. It's a rather fragile part of the anatomy, but **if it's kept straight** then it's much less likely to get hurt. If you're in your twenties, and many readers are likely to be, you're probably going to laugh at the thought of straining your back. But a little later in life – and not all that much later – people stop laughing! Almost any doctor can tell of patients hobbling into the surgery in the Spring. They've not exercised their backs much in Winter, but with better weather they're out digging gardens – and straining their backs.

So, how do you avoid hurting your back? It's very important indeed when lifting something at all heavy to keep the back as straight as possible. The British Safety Council says:

1. Stand as close to the object as possible and spread your feet to form a stable base.
2. Bend your knees, keeping your back in a straight line – but don't bend your knees too much.
3. Grasp the object and raise your head as you start to lift.
4. Lift, using your legs, not your back.

But what if you can't follow that procedure – the thing's too heavy or too awkward a shape? Then you get someone to help. It sometimes needs a bit of courage to go and ask a friend or colleague to assist you in lifting something, but it's worth it.

Razor blades

Essential for editing tape, these are becoming less common with the increased use of digital editing, but they're still around. I don't need to list the possible risks arising from accidental cuts. **It's very important that used blades are got rid of safely.** Strong plastic boxes with a slot for the blades are found in many studios – they can be bought from suppliers of audio equipment, but it wouldn't be difficult to make a safe receptacle.

Fire

Everyone ought to know what to do if fire breaks out. Sound studios are usually fairly safe areas, provided there aren't idiots putting lighted cigarette ends into waste paper bins. Nevertheless, ALWAYS BE PREPARED FOR THE WORST.

There are a few basic – and really very obvious – rules:

1. Know where the fire exits are, and make sure they're not obstructed.
2. Know how to call the fire brigade.
3. Be able to identify the correct fire extinguisher.

Knowing what extinguisher to use on what fire is obviously vital. In any studio area, if there's a fire it's likely to be an electrical one – and you DON'T use water to put that kind of fire out! Water is, of course, a conductor of electricity, so putting water on an electrical fire will make things worse. Carbon Dioxide (CO_2) extinguishers (colour code BLACK) are generally satisfactory for electrical fires. It's best, though, to have a chart on a wall, showing what extinguisher to use when. Such charts aren't difficult to get hold of.

Fire doors

These are filled with special material that makes them 'fire-retardant'. That means that it takes quite a while, say half an hour, for heat from a fire to penetrate through the door. Obviously, then, these doors in corridors can help to confine a fire and give people a better chance of leaving the building safely. But fire doors must be kept closed! Don't wedge them open with fire extinguishers, which is a very bad common practice!

And finally

With any event where the public is present, it is advisable to check on insurance coverage. Don't assume that any accidents occurring as a result of your activities, whether you are to blame or not, are automatically covered.

There is an almost endless list of Safety Do's and Don'ts. However, the single most important factor is the right attitude of mind! If you always think Safety you're almost always going to be working safely.

SAFE THINKING = SAFE WORKING!

Questions

1. A small sound mixer is rated at 750 W electrical consumption. What fuse would be appropriate in the mains plug?
   a. 1 A   b. 3 A   c. 13 A

2. A ladder used to get on to a scaffolding should be fixed at an angle of about
   a. 40°   b. 60°   c. 75°   d. 85°

3. A mains plug has accidentally got wet. What action would be most sensible?
   a. Open it up and wipe it dry as far as possible before using it.
   b. Open it up and wipe it dry as far as possible. Then put it in a warm place for a few hours.
   c. Open it up and wipe it dry as far as possible. Then put it in a warm place for a few hours. Finally, inspect it carefully for signs of moisture.
   d. Reject it.

4. What is the colour code for carbon dioxide fire extinguishers?
   a. Red   b. Black   c. Blue   d. Green

Copyright

The laws relating to copyright are very complicated. Basically, it is illegal to copy any recording whether for private use or not. Some recordings are described as 'Public Domain', which implies that they are free from any copyright restrictions. This is not necessarily the case: the actual recording may have gone out of copyright but the rights to the work (e.g. a piece of music) may still be within the 50 years' protection after the composer's death, if the composer is no longer alive.

The addresses of the organizations which should be consulted where there is any doubt are given below:

Mechanical-Copyright Protection Society Ltd (MCPS)
Elgar House, 41 Streatham High Road, Streatham, London SW16 1ER. (Protects the interests of composers and arrangers and issues licences to recording organizations.)

Performing Right Society (PRS)
29–33 Berners Street, London W1P 4AA. (Issues licences for any kind of public performance, including TV and radio.)

Phonographic Performance Ltd (PPL)
Ganton House, 14–22 Ganton Street, London W1V 1LB. (Similar in licence issue to the Performing Right Society.)

Miscellaneous data

Resistance of copper conductors (approximate)

SWG no.	Diameter (mm)	Cross-sectional area (mm²)	Resistance/m (Ω)
10	3.25	8.3	0.002
12	2.64	5.48	0.003
14	2.03	3.24	0.005
16	1.62	2.08	0.008
18	1.22	1.17	0.015
20	0.91	0.66	0.026
22	0.71	0.40	0.043
24	0.56	0.25	0.070
26	0.46	0.16	0.10
28	0.38	0.11	0.15
30	0.32	0.08	0.22

Standard prefixes for multiples and submultiples

Multiplication		Prefix	Symbol
1 000 000 000	$= 10^9$	giga	G
1 000 000	$= 10^6$	mega	M
1000	$= 10^3$	kilo	k
1/1000	$= 10^{-1}$	milli	m
1/1 000 000	$= 10^{-6}$	micro	μ
1/1000 000 000	$= 10^{-9}$	nano	n

Decibel equivalents for power and voltage ratios

Decibels	Power ratio	Voltage ratio
0	1	1
1	1.26	1.12
2	1.58	1.26
3	2.0	1.41
4	2.51	1.58
5	3.16	1.78
6	3.98	2.0
7	5.01	2.24
8	6.31	2.51
9	7.94	2.82
10	10.0	3.16
20	100.0	10.0
30	1000	31.6
40	10 000	100
50	100 000	316
60	1 000 000	1000

Conversion factors

1 inch	=	2.54 cm
1 foot	=	30.48 cm
1 yard	=	91.4 cm
1 mile	=	1609 metres (m)
1 cm	=	0.394 inch
1 m	=	39.4 inches
1 km	=	0.62 miles
1 m/s	=	2.237 mph
1 mph	=	0.447 m/s
	=	1.609 km/h
1 kgm	=	2.204 lb
1 pascal (Pa)	=	1.450×10^{-4} lb/inch^2
1 atmosphere	=	1.013×10^{-5} Pa
1 millibar	=	102 Pa

Connectors

It is probably good practice to assume initially that any audio connectors one comes across are not inevitably wired correctly! There can be some deviations from convention. Within one's own system, consistency of connection is clearly very important, otherwise out-of-phase conditions may occur, hum may intrude and there might even be a total loss of signal. The tables below and on the next pages give the generally accepted conventions for plug and socket wiring. These are given as pin numbers rather than as diagrams, as the latter can sometimes cause confusion: it is easy to forget, for example, that a diagram may represent the front of a plug where one solders the wires on at the back!

Balanced circuits

Connector type	Signal+	Signal–	Earth (screen)
XLR style	Pin 2*	Pin 3*	Pin 1
3-terminal jack (e.g. PO style or so-called 'stereo' jack, 6.3 mm or miniature, wired for a single balanced circuit)	Tip	Ring	Sleeve
3-pin DIN (as fitted to some microphones)	Pin 1	Pin 3	Pin 2

*XLR pins 2 and 3 are reversed by some organizations and in some, mainly American, microphones. When connected to equipment using the other standard, this can cause a phase reversal in balanced circuits and a possible loss of signal in unbalanced ones.

Unbalanced circuits

Connector type	Signal+	Earth (screen) and signal–
XLR style	Pin 2*	Pins 3* and 1
Phono	Pin	Sleeve
2-terminal (6.3 mm or miniature)	Tip	Sleeve
3-terminal jack (6.3 mm or miniature)	Tip	Ring and sleeve
3-terminal jack (6.3 mm or miniature) wired for stereo	L: Tip R: Ring	Sleeve

*See note on previous table.

DIN connectors

Note that there can be variations depending on the manufacturer of the equipment.

DIN 180° (5-pin) (Stereo: L = Left, R = Right)

Microphones	L Pin 1 R Pin 4	Pin 2	Pins 3 and 5 may carry the polarizing voltage
Tape recorder inputs	L Pin 1 R Pin 4	Pin 2	
Tape recorder outputs (Low impedance)	L Pin 1 R Pin 4	Pin 2	
(High impedance)	L Pin 3 R Pin 5	Pin 2	

Further reading

There is a wide range of printed matter devoted to the topics dealt with in this book. In a few cases, which obviously cannot be named, what is written has to be treated with caution as there are sometimes incorrect or dubious statements. The list below can be recommended, though. We must add that the list is far from complete. The absence of a book from the list may not mean that it is not a good one.

To help the reader, the books are rated as 'Easy', 'Moderate' or 'Difficult':

Easy means that anyone who has coped with the present book should have little or no difficulty in reading the one listed.

Moderate implies that the work is a little more advanced than this book, but should be seen as a possible next step forward for the interested reader.

Hard should be taken as much more advanced. It would be best to look at a copy in a library before deciding whether or not to buy it.

Borwick, John. *Microphones*, Focal Press, 1990. An up-to-date and authoritative book. *Moderate*.

Collons, Martin. *High Performance Loudspeakers*, Pertech Press. First published 1985. *Moderate*.

Derry, Roger. *PC Audio Editing*, Focal Press, March 2000. *Easy to moderate*.

Johnston, Ian. *Measured Tones*, Adam Hilger/IOP Publishing, 1989. A very interesting and readable account of the science of musical instruments. *Easy*.

McLeish, Robert. *Radio Production*, 2nd Edition, Focal Press, 1994. A good outline of the non-technical side of the production of sound-only programmes. *Easy*.

Moore, Brian C. J. *An Introduction to the Psychology of Hearing*, Academic Press, 1989. Rather advanced but very interesting to those who want to find out more about this intriguing subject. *Hard*.

Nisbett, Alec. *The Use of Microphones*, 3rd Edition, Focal Press, 1989. A comprehensive reference book. *Easy*.

Rumsey, Francis. *Digital Audio Operations*, Focal Press, 1991. A fairly advanced book but it can be recommended to anyone who really wants to delve into digits. *Hard*.

Rumsey, Francis. *MIDI Systems and Control*, Focal Press, 1990. Not so much for the musician who uses MIDI as for the person who wants to understand what is going on. *Moderate*.

Talbot-Smith, Michael (ed.). *Sound Engineer's Pocket Book*, Focal Press, 2001. Exactly what it says. A collection of data useful to the professional or serious amateur. Cannot be rated as *Hard* or *Easy*!

In addition there are magazines. Most are directed towards promoting the sales of either equipment or recordings and only a few towards techniques. Many, however, are often worth looking at, but the reader should look at sample copies carefully before deciding to take out a subscription!

Answers

(You can find more questions at http://www.cwc.ac.uk/sound)

Answers to the questions at the end of Chapter 1
1. c. 16–16 000 Hz
2. b. 340 m/s
3. c. Above roughly 700 Hz

Answers to the questions at the end of Chapter 2
1. b. Two sound pressures   e. Two powers
2. c. 400 Hz
3. c. 100 Hz

Answers to the questions at the end of Chapter 3
1. b. A minimum of vibration
2. c. 60 dB
3. b. 0.5 s
4. a. 1 square metre of 100% absorber

Answers to the questions at the end of Chapter 4, Part 1
1. b. They are generally robust and reliable, and
   c. The quality of their output, while good, is likely to be inferior to some other types
2. a. They are usually fragile, and
   b. Very good quality is possible because of the lightness of the ribbon
3. c. It is like two spheres in contact
4. b. 25–30 dB
5. c. 45° on each side of the rear

Answers to the questions at the end of Chapter 4, Part 2
1. b. Standard three-core microphone cable
2. d. Electrostatic (but personal microphones are powered by batteries)

Answers to the questions at the end of Chapter 5
1. b, c and d (all except omnidirectional)
2. b. Looking at the instrument in the same direction that an audience would do

Answers to the questions at the end of Chapter 6, Part 1
1. c. Average signal voltages
2. d. Peak signal voltages
3. c. 1–2 seconds

Answers to the questions at the end of Chapter 6, Part 2
1. b. 6
2. c. 1–2%

Answers to the questions at the end of Chapter 7, Part 1
1. d. 1 ms
2. b. Microphones too widely spaced
3. a. Fully right
4. a. An out-of-phase condition

Answers to the questions at the end of Chapter 7, Part 2
1. c. 180°
2. b. 135°
3. a. 90°
4. Coincident omni microphones do not give stereo!
5. c. Figures of eight (d is a silly answer!)

Answers to the questions at the end of Chapter 8, Part 1
1. c. To allow the fader to be set to an optimum position
2. b. ϕ indicates a phase reverse control

Answers to the questions at the end of Chapter 8, Part 2
1. a. With XLR sockets in the studio (a microphone with XLR connectors has a PLUG on it)
2. d. Accurate 0.775 V, about 1 kHz (sometimes the frequency is an accurate 1 kHz, but it is the voltage which MUST be accurate)

Answers to the questions at the end of Chapter 9
1. c.   The input rises 3 dB for each 1 dB rise in the output above the threshold
2. c.   20:1 or more

Answers to the questions at the end of Chapter 10, Part 1
1. a.   44 000 (44 100 to be precise)
2. c.   9
3. d.   Musical Instrument Digital Interface

Answers to the questions at the end of Chapter 11, Part 1
1. d.   Erase, record, replay
2. a.   4.75 cm/s
3. a.   Reduced high frequencies

Answers to the questions at the end of Chapter 12
1. a.   Wide horizontally, narrow vertically
2. c.   Cardioid

Answers to the questions at the end of Chapter 14
1. c.   13 A. The mixer would probably take only about 3 A normally but there could be a surge when switching on
2. c.   About 75°
3. d.   Reject is probably the best answer, but circumstances might not allow this, in which case the next best answer is c – open it up and wipe it dry as far as possible. Then put it in a warm place for a few hours before inspecting it for signs of moisture
4. b.   Black

Index

100-volt systems, 168

A-B comparisons, 20, 21, 80
Absorption coefficient, 36, 37
Acoustics, 25
ADC, 143
Air particle displacement, 30
Airborne sound, 26
Amplitude, 2
Analogue recording, 151
Angle of acceptance, 98
Answers, 191
Antinode, 30
Anti-vibration mounts, 28
Artificial reverberation, 140
Attenuator, 107
Aural monitoring, 71
Auxiliary outputs, 106
AVM, *see* Anti-vibration mounts

Balance, 66
Balanced wiring, 58, 113
Bar, 8, 12
Barometric pressure, 1
Bass:
    cut/lift, 110, 116
    tip-up, 54
Binary arithmetic, 136
Binaural stereo, 101
Bit rate, 138
Block diagram, 105
Boundary microphone, 48, 55

Cables, loudspeaker, 86
Capsule, microphone, 51

Carbon granule microphone, 42
Cardioid microphone, 46, 54
Cassette, quality, 153
CDs, 145
    cleaning, 148
Channel, 105
    fader, 108
    sensitivity control, 106, 107
Circuit breakers, 177
Cocktail party effect, 20
Coincident pair, 90
Column loudspeaker, 166
Compact disc, *see* CDs
Compression, 1
    ratio, 128
Connectors, 187, 188
Control:
    electronic, 127
    levels, 123
    manual, 111
Conversion factors, 186
Copyright, 184
Crossover unit, 79
Crystal microphone, 42
Cycles per second, 2

DAC, 143
DAT, 157
Data:
    compression, 143
    miscellaneous, 185
dB, *see* Decibel
dB(A), 16
Decay time, 129

Decibel, 10, 13
  equivalents, 186
Delay, digital, 140
DI box, *see* Direct injection, 69
Diaphragm (microphone), 40
Diffraction, 4, 5
Digital:
  audio, 133
  recording, 156
Direct:
  injection , 69
  sound, 63
Discussions, 68
Distortion, overload, 61
Diversity reception, 60
Dolby® systems, 154
Drive unit, 77
Dynamic:
  microphone, 41
  range, 71

Ear, response, 14
Editing:
  digital, 162
  general, 160
  MiniDisc, 163
  tape, 162
Electret microphone, 51
Electric shock, 178
Electrical safety, 175
Electrostatic microphone, 42, 51
Enclosure, 77
EQ, 105, 109, 116
Equal tempered scale, 17
Equalization, *see* EQ
Error:
  correction, 147, 150
  detection, 139
Expanders, 131

Fader, 105
False bass, 23
FB, *see* Foldback
Figure-of eight microphone, 45, 52
Filters, 118
Fire, 181
  doors, 182
  extinguishers, 182
Foldback, 119

Frequency, 3
  ranges, 22
Fundamental, 18
Fuses, 175

Graphic equalizers, 118
Gun microphone, 47, 55

Haas effect, 20
Harmonics, 18
Headphones, 80, 102
Heads, erase, record, replay,
  152
Heavy weights, lifting, 180
Helmholtz resonator, 78, 84
Hertz, 2
High-level source, 107
Hole-in-the middle stereo, 89
Howl-rounds, 165, 167
Hypercardioid microphone, 46,
  55

Indirect sound, 63
Initial time delay, 33
Inputs, mixer, 113
Insert point, 106
Intensity, 8
Inter-channel differences, 90
Internal balance, 68
International A, 16
Interviews, 66
Inverse square law, 9
Isolating transformers, 177

Keyboard, 17

Ladders, 180
Lambda, 2
LED level indicators, 74
Limiting, 128
Line-source loudspeaker, 166
Longitudinal waves, 1
Loudness, 15, 17
Loudspeaker,
  cables, 86
  matching, 94
  power, 85
Loudspeakers, 75
Low-level source, 107

Mass law, 35
Mechanical safety, 179
Microphone:
  techniques, stereo, 97
  transducer comparison, 43
Microphones, 40
Microwatt, 7
Middle C, 19
Mid-frequency unit, 79
MIDI, 142
MiniDisc®, 157
Mixers, sound, 104
Monitoring, 71, 82
Moving coil:
  loudspeaker, 76
  transducer, 40
MP3, 159
Multi-miking, 97
Music, 170
Musical:
  groups, small, 69
  notes, frequencies, 19
  quality, 18

Near-field monitor, 79
Newton, 8
NICAM, 148
Node, 30
Noise:
  and hearing, 178
  gates, 131
  reduction (tape), 154

Obstacles, 4
Octave, 17
Omnidirectional microphone, 44,
  52
Optics of a CD system, 146
Output, mixer, 111

PA, 119, 164
  indoor, 165
  outdoor, 165
Panpot, 91
Parametric equalizers, 119
Parity, 139
Pascal, 8
Peak Programme Meter, see PPM
Period, 2

Personal microphone, 48
Phantom power, 57
Phase, 94
  reverse, 108
Phase-shift microphone, 54
Piano, 67
Ping-pong stereo, 89
Pitch, 16, 22
  changing, 141
PO jack, 114
Polar:
  diagram, 44
  responses, 44
Porous absorber, 36
Power, 6, 7
PPM, 73, 83
Presence, 110, 117
Presenter, solo, 67
Pressure, 8
  gradient, 52
  operated microphone, 52
  zone microphone, 48
Proximity effect, 54
Public Address, see PA

Quantizing, 135
Quarter-inch tape, 151

r.f. electrostatic microphone, 52
Radio microphone, 48, 59
Rarefaction, 1
Razor blades, 181
Recovery time, 129
Reference voltages, 82
Regeneration of pulses, 137
Resistance of copper conductors, 185
Reverberation, 31
  time, 31, 38
Ribbon microphone, 41
RT, see Reverberation, time
Rumble, 62

Sabine, 37
Sabine's formula, 38
Sampling, 134
Sealed enclosure loudspeaker, 78
Semitone, 17
Sensitivities of microphones, 56
Signal-to-noise ratio, 72

Singing group, 68
Solid state recording, 158
Sound:
  absorption, 35
  effects, 171
  isolation, 25, 35
  level meters, 16
  reinforcement, 164
Sound-in syncs, 149
Standard prefixes, 185
Standing waves, 29, 30
Star quad, 58
Starting transients, 18
Stereo, 88
  listening, 92
Structure-borne sound, 27
Symbols, 118, 121

Talkback , 121
Tape:
  cleanliness, 155
  tracks, 154
Terminology, stereo, 91
Threshold, 128
  of hearing, 14, 15
Timbre, 18

Time-of-arrival difference, 88
Top (treble) cut/lift, 110, 117
Tracking, CD, 146
Transducer (microphone), 40
Transients, starting, 18
Transverse waves, 2
Tweeter, 79

Unbalanced wiring, 113
Units used in sound, 6

VCA, 127
Velocity of sound waves, 3, 11
Vented enclosure, 78, 79
Visual monitoring, 71
Voltage controlled amplifier,
  see VCA
VU meter, 73, 83

Wavelength, 2, 3
Wideband porous absorber, 36
Windshields, 62

XLR plugs, sockets, 114

Zero level, 82§

Focal Press

www.focalpress.com

Join Focal Press on-line
As a member you will enjoy the following benefits:
- an email bulletin with **information on new books**
- a regular **Focal Press Newsletter**:
  - o   featuring a selection of new titles
  - o   keeps you informed of **special offers, discounts and freebies**
  - o   alerts you to **Focal Press news and events** such as author signings and seminars
- complete access to **free content** and reference material on the focalpress site, such as the focalXtra articles and commentary from our authors
- a **Sneak Preview** of selected titles (sample chapters) *before* they publish
- a chance to have your say on our **discussion boards** and **review books** for other Focal readers

Focal Club Members are invited to give us feedback on our products and services. Email: worldmarketing@focalpress.com – we want to hear your views!

Membership is FREE. To join, visit our website and register. If you require any further information regarding the on-line club please contact:

> Emma Hales, Marketing Manager
> Email: emma.hales@repp.co.uk
> Tel: +44 (0) 1865 314556
> Fax: +44 (0)1865 315472
> Address: Focal Press, Linacre House,
> Jordan Hill, Oxford, UK, OX2 8DP

Catalogue
For information on all Focal Press titles, our full catalogue is available online at www.focalpress.com and all titles can be purchased here via secure online ordering, or contact us for a free printed version:

USA
Email: christine.degon@bhusa.com

Europe and rest of world
Email: jo.coleman@repp.co.uk
Tel: +44 (0)1865 314220

Potential authors
If you have an idea for a book, please get in touch:

USA
Lilly Roberts, Editorial Assistant
Email: lilly.roberts@bhusa.com
Tel: +1 781 904 2639
Fax: +1 781 904 2640

Europe and rest of world
Christina Donaldson, Editorial Assistant
Email: christina.donaldson@repp.co.uk
Tel: +44 (0)1865 314027
Fax: +44 (0)1865 314572